Edexcel A2 Biology Revision Guide

for SNAB and concept-led approaches

REVISION GUIDE

Gary Skinner

Robin Harbord

Ed Lees

A PEARSON COMPANY

Published by Pearson Education Limited, a company incorporated in England and Wales, having its registered office at Edinburgh Gate, Harlow, Essex, CM20 2JE. Registered company number: 872828

Edexcel is a registered trade mark of Edexcel Limited

First published 2009

12 11 10
10 9 8 7 6 5 4 3 2

British Library Cataloguing in Publication Data
A catalogue record for this book is available from the British Library

ISBN 978 1 846905 99 5

External project management by Sue Kearsey
Edited by Liz Jones
Typeset by 320 Design Ltd
Illustrated by Oxford Designers & Illustrators
Cover photo © Jupiter Unlimited
Printed in Malaysia (CTP-VVP)

Acknowledgements
Edexcel review by Martin Furness-Smith, UYSEG review by Anne Scott
We would like to thank Damian Riddle, Anne Scott and Elizabeth Swinbank for their contributions

Disclaimer
This material has been published on behalf of Edexcel and offers high-quality support for the delivery of Edexcel qualifications.

This does not mean that the material is essential to achieve any Edexcel qualification, nor does it mean that it is the only suitable material available to support any Edexcel qualification. Edexcel material will not be used verbatim in setting any Edexcel examination or assessment. Any resource lists produced by Edexcel shall include this and other appropriate resources.

Copies of official specifications for all Edexcel qualifications may be found on the Edexcel website: www.edexcel.com

Contents

Revision techniques

Getting started can be the hardest part of revision, but don't leave it too late. Revise little and often! Don't spend too long on any one section, but revisit it several times, and if there is something you don't understand, ask your teacher for help.

Just reading through your notes is not enough. Take an active approach using some of the revision techniques suggested below.

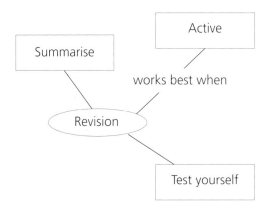

Summarising key ideas

Make sure you don't end up just copying out your notes in full. Use some of these techniques to produce condensed notes.

- Tables and lists to present information concisely
- Index cards to record the most important points for each section
- Flow charts to identify steps in a process
- Diagrams to present information visually
- Spider diagrams, mind maps and concept maps to show the links between ideas
- Mnemonics to help you remember lists
- Glossaries to make sure you know clear definitions of key terms

Include page references to your notes or textbook. Use colour and highlighting to pick out key terms.

Check the spec.
If you use resources from elsewhere, make sure they cover the right content at the right level.

Active techniques

Using a variety of approaches will prevent your revision becoming boring and will make more of the ideas stick. Here are some methods to try.

- Explain ideas to a partner and ask each other questions.
- Make a podcast and play it back to yourself.
- Use PowerPoint to make interactive notes and tests.
- Search the internet for animations, tests and tutorials that you can use.
- Work in a group to create and use games and quizzes.

Test yourself

Once you have revised a topic, you need to check that you can remember and apply what you have learnt.

- Use the questions from your textbook and this revision guide.
- Get someone to test you on key points.
- Try some past exam questions.

How to use this Revision Guide

Welcome to your **Edexcel A2 Biology Revision Guide**, perfect whether you're studying Salters Nuffield Advanced Biology (the orange book), or the 'concept-led' approach to Edexcel Biology (the green book).

This unique guide provides you with tailored support, written by senior examiners. They draw on real 'ResultsPlus' exam data from past A-level exams, and have used this to identify common pitfalls that have caught out other students, and areas on which to focus your revision. As you work your way through the topics, look out for the following features throughout the text.

ResultsPlus Examiner's Tip
These sections help you perform to your best in the exam by highlighting key terms and information, analysing the questions you may be asked, and showing how to approach answering them. All of this is based on data from real-life A-level students!

ResultsPlus Watch Out
The examiners have looked back at data from previous exams to find the common pitfalls and mistakes made by students – and guide you on how to avoid repeating them in *your* exam.

Quick Questions
Use these questions as a quick recap to test your knowledge as you progress.

Thinking Task
These sections provide further research or analysis tasks to develop your understanding and help you revise.

Worked Example
The examiners guide you through complex equations and concepts, providing step-by-step guidance on how to tackle exam questions.

Each topic also ends with:

Topic Checklist
This summarises what you should know for this topic, which specification point each checkpoint covers and where in the guide you can revise it. Use it to record your progress as you revise.

ResultsPlus Build Better Answers
Here you will find sample exam questions with exemplar answers, examiner tips and a commentary comparing both a basic and an excellent response; so you can see how to get the highest marks.

Practice Questions
Exam-style questions, including multiple-choice, offer plenty of practice ahead of the exam.

Both Unit 4 and Unit 5 conclude with a **Specimen Paper** to test your learning. These are not intended as timed, full-length papers, but provide a range of exam-style practice questions covering the range of content likely to be encountered within the exam.

The final unit consists of advice and support on research skills, giving guidance on Practical Assessment to help you write better individual investigations.

Answers to all the in-text questions, as well as detailed, mark-by-mark answers to the specimen papers, can be found at the back of the book.

We hope you find this guide invaluable. Best of luck!

Question types in GCE Biology

Multiple choice

A good multiple choice question gives you the correct answer and other possible answers which seem plausible.

Triglycerides are composed of: (1)
3 glycerol molecules and 3 fatty acid molecules ☐
1 glycerol molecule and 3 fatty acid molecules ☐
1 glycerol molecule and 1 fatty acid molecule ☐
3 glycerol molecules and 1 fatty acid molecule ☐

The best way to answer a multiple choice question is to read the question and try to answer it before looking at the possible answers. If the answer you thought of is amongst the possible answers – job done! Just have a look at the other possibilities to convince yourself that you were right.

If the answer you thought of isn't there, look at the possible answers and try to eliminate wrong answers until you are left with the correct one.

You don't lose any marks by having a guess (if you can't work out the answer) – remember you won't score anything by leaving the answer blank! If you narrow down the number of possible answers, the chances of having a lucky guess at the right answer will increase.

To indicate the correct answer, put a cross in the box following the correct statement. If you change your mind, put a line through the cross and fill in your new answer with a cross.

How Science Works

The idea behind How Science Works is to give you insight into the ways in which scientists work: how an experiment is designed, how theories and models are put together, how data are analysed, how scientists respond to factors such as ethics, and so on, and the way society is involved in making decisions about science.

Many of the HSW criteria are practical and will be tested as part of your practical work. However, there will be questions on the written unit papers that cover some HSW criteria. Some of these questions will involve data or graph interpretation (HSW 5) – see the next section.

The other common type of HSW question will be based on the core practicals. Questions will concentrate not so much on what you did, but why various steps in the core practical were important. It's important, therefore, that you know what the various steps in each core practical were designed to do; and that you revise the core practicals.

For example, think about the questions that could be asked in a Unit 2 paper on the core practical in Topic 3: 'Describe the stages of mitosis and how to prepare and stain a root tip squash in order to observe them practically.' Here, suitable questions could include:
- why do we use only the tip of the root?
- which stain do we use?
- why do we place the cut tip in acid before staining?
- what safety precautions would be relevant here, and why?

You'll also commonly get asked questions involving designing an investigation: these are likely to involve pieces of familiar practical work. The **CORMS** prompt may be useful here:

Control	– Are you investigating simply with / without a particular factor?
	What range of values are you looking at?
Organism	– Are you using organisms of the same sex / age / size / species?
Repeat	– Take readings more than once and average
Measure	– What are you measuring? How will you measure this? What units?
Same	– Which variable(s) are you keeping constant?

Other HSW questions may concentrate on ethical issues surrounding topics such as gene therapy or GM foods.

Interpretation of graphs

The graph below shows the results of a survey in America, on the incidence of heart disease in adults aged 18 and older.

> Using the information in the graph, describe how the incidence of heart disease is affected by age and gender. (3)

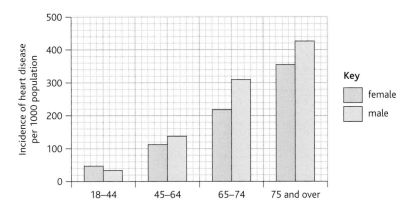

You almost always get one mark for stating the general trend – in this case that the incidence of heart rate increases, with increasing age, for both genders. You can then concentrate on individual aspects of the data. In this case, what stands out most is that females have a lower incidence of heart disease than males, except in the 18–44 age group. You may also comment that the difference between the incidence in males and females is largest in the 65–74 age group.

Finally, there is always one mark for manipulation of data. Note that this must be *manipulation* – you don't get marks for reading off the graph and stating the numbers, you have to do something with them!

Extended questions

In the A2 units (Topic 4 and Topic 5) you will come across questions with larger numbers of marks, perhaps up to 6 or 7 marks in the question.

Questions in these units are designed to be synoptic – in other words, they are designed for you to show knowledge gained in earlier units. Bear this in mind when you answer the question: try to include relevant knowledge from your AS course when answering these questions.

Remember, too, that if the question is worth 6 marks, you need to make six creditworthy points. Think about the points that you will make and put them together in a logical sequence when you write your answer. On longer questions, the examiners will be looking at your QWC (Quality of Written Communication) as well as the answer you give.

Photosynthesis 1

Photosynthesis involves the **reduction** of carbon dioxide (CO_2) to carbohydrate. These carbohydrates can be used to provide energy in respiration (see Topic 7). The hydrogen for this process comes from the splitting of water by light, the waste oxygen being released into the atmosphere.

This can be summarised like this:

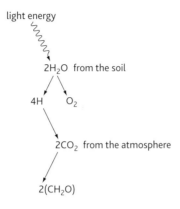

Diagram showing the main steps in photosynthesis. Light energy splits strong bonds in water to give hydrogen and oxygen (which is released to the atmosphere). The hydrogen is stored in a fuel (glucose) by reducing carbon dioxide to carbohydrate.

If you imagine the above happening three times, you have $C_6H_{12}O_6$, which is glucose (and a number of other sugars, see Topic 1.1 AS).

So, the overall equation is:

$$6CO_2 + 6H_2O \xrightarrow[\text{chlorophyll}]{\text{light}} C_6H_{12}O_6 + 6O_2$$

The splitting of water by light is called **photolysis**. The energy for this step is first trapped by a pigment molecule called chlorophyll.

The overall process is achieved in two linked stages called the **light-dependent reactions** and the **light-independent reactions**.

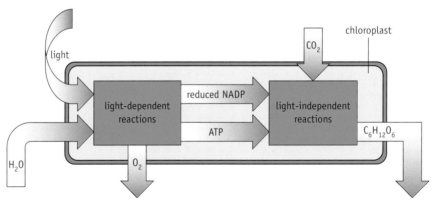

Summary of photosynthesis showing the linked light-dependent reactions (in which water is split and ATP and reduced NADP are made), and the light-independent reactions (in which the energy from ATP and reducing power from reduced NADP are used to make sugar).

The light-dependent reactions

In this process, a pair of electrons from **chlorophyll** is boosted to a higher energy level by the light energy it has trapped. Here they are accepted by an electron acceptor and then passed along a chain of carriers. Energy released is used to convert ADP and inorganic phosphate (Pi) into ATP. This process is called **photophosphorylation** (the light-driven addition of phosphate). The electrons then enter another chlorophyll molecule. The electrons eventually pass to NADP with the hydrogen from water to form reduced NADP. The ATP and reduced NADP are then used in the light-independent reactions to make carbohydrate from carbon dioxide.

Key

→ flow of electrons in non-cyclic photophosphorylation

- ▸ flow of electrons in cyclic photophosphorylation

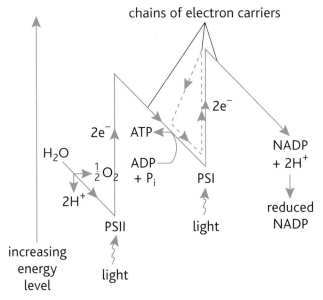

Diagram showing light-dependent reactions. Photosystems PSI and PSII are the two special chlorophyll molecules which can release their electrons when struck by light.

ResultsPlus
Examiner tip

You will not be expected to recall details of photosystems but you may be expected to understand their roles if they are part of an exam question.

ResultsPlus
Examiner tip

Never try to answer a question with just a diagram, always include some explanation of what is going on.

☼☼ Thinking Task

Q1 The diagram shows the results of an experiment in which air containing isotopes $^{18}O_2$ and water containing $H_2{}^{16}O$ was bubbled through a suspension of algae.

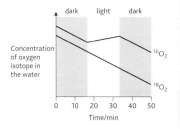

Say how these data support the idea that the oxygen given off by plants in photosynthesis comes from water.

? Quick Questions

Q1 Name the substance which provides reducing power (electrons) and the one which supplies energy for the light-independent reactions of photosynthesis.

Q2 What is:
a reduction
b oxidation?

Photosynthesis 2

The light-dependent reactions make **ATP** and **reduced NADP** which are then used in the light-independent reactions (Calvin cycle). The reduced NADP provides **reducing power** (electrons or hydrogen) and the ATP provides the energy for the process of making carbon dioxide into carbohydrate.

The key steps in the **Calvin cycle** are shown in the diagram.

1 Carbon dioxide combines with a 5-carbon compound called **ribulose bisphosphate** (**RuBP**). This reaction is catalysed by the enzyme **ribulose bisphosphate carboxylase** (**RuBISCO**), the most abundant enzyme in the world.

2 The 6-carbon compound formed is unstable and immediately breaks down into two 3-carbon molecules, **glycerate 3-phosphate** (**GP**).

3 This 3-carbon compound is reduced to form a 3-carbon sugar phosphate called **glyceraldehyde 3-phosphate** (**GALP**). The hydrogen for the reduction comes from the reduced NADP from the light-dependent reactions. ATP from the light-dependent reactions provides the energy required for the reaction.

5 Ten out of every 12 GALPs are involved in the recreation of RuBP. The ten GALP molecules rearrange to form six 5-carbon compounds; then phosphorylation using ATP forms RuBP.

4 Two out of every 12 GALPs formed are involved in the creation of a 6-carbon sugar (hexose) which can be converted to other organic compounds, for example amino acids or lipids.

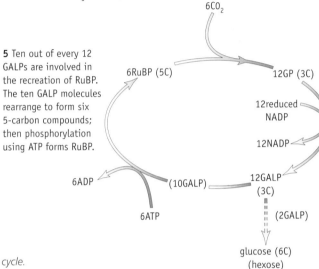

Outline of the Calvin cycle.

ATP

Adenosine triphosphate (ATP) provides energy for chemical reactions in the cell. When energy is needed, phosphate is removed from the ATP to give ADP and a phosphate. The energy is released when the phosphate forms bonds with water. In the photosynthesis light-dependent reactions, ATP is made using energy from light.

$$ATP \longrightarrow ADP + P_i + energy$$

In photosynthesis, the ATP made is used as a source of energy in the light-independent reactions. ATP is also used widely in organisms as a way of transferring energy. It is an intermediate between energy-producing reactions and those that need energy.

Some of the glucose made in the Calvin cycle is used by the plant in respiration. The rest is used to synthesise all the molecules on which the plant relies, for example other simple sugars, polysaccharides, amino acids, lipids and nucleic acids.

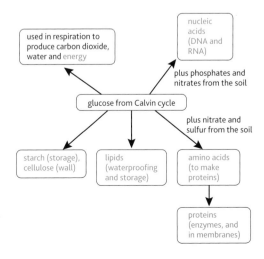

Diagram showing some of the fates of the glucose made in photosynthesis.

Where does photosynthesis happen?

In all eukaryotic cells there are membrane-bound structures called **organelles**. These are the sites of specialised processes within the cell. For photosynthesis, plant cells have a structure called the **chloroplast**. The diagram shows the functions which each part carries out.

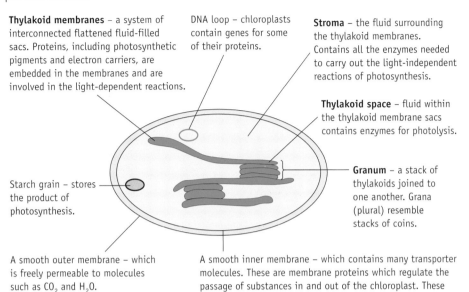

Thylakoid membranes – a system of interconnected flattened fluid-filled sacs. Proteins, including photosynthetic pigments and electron carriers, are embedded in the membranes and are involved in the light-dependent reactions.

DNA loop – chloroplasts contain genes for some of their proteins.

Stroma – the fluid surrounding the thylakoid membranes. Contains all the enzymes needed to carry out the light-independent reactions of photosynthesis.

Thylakoid space – fluid within the thylakoid membrane sacs contains enzymes for photolysis.

Granum – a stack of thylakoids joined to one another. Grana (plural) resemble stacks of coins.

Starch grain – stores the product of photosynthesis.

A smooth outer membrane – which is freely permeable to molecules such as CO_2 and H_2O.

A smooth inner membrane – which contains many transporter molecules. These are membrane proteins which regulate the passage of substances in and out of the chloroplast. These substances include sugars and proteins synthesised in the cytoplasm of the cell but used within the chloroplast.

? Quick Questions

Q1 There are two steps where ATP is used in the Calvin cycle. Where are they?

Q2 Whereabouts in the Calvin cycle is RUBISCO used and what does it do?

Q3 Copy and complete the table of chloroplast functions below by suggesting, for each structure within the chloroplast, the features that are adapted for these functions.

Structure	Function(s)	Features
thylakoid membrane	light-dependent reactions	
thylakoid space	photolysis of water	
granum	provides a site for light-dependent reactions	
stroma	light-independent reactions	
outer membrane	fully permeable	
inner membrane	permeable to many substances which need to enter or leave the chloroplast	

⚙ Thinking Task

Q1 Look at the diagram of the Calvin cycle on page 10. Work out how many carbon atoms are involved at each stage (RuBP, CO_2, GP, GALP, glucose).

Energy transfer, abundance and distribution

Plants make glucose in photosynthesis. This can be turned into other molecules including starch, cellulose, proteins and fats. This biomass is food for both humans and every other living thing on Earth, including the plants themselves.

The rate at which energy is incorporated into organic molecules in the plants in photosynthesis is called **gross primary productivity** (**GPP**). Plants use some of the organic molecules in respiration (see Topic 7). If we find out the figure for GPP and take away the amount of energy used in respiration (R), what is left is the rate at which energy is transferred into new plant biomass that can be eaten by herbivores or decomposers. This is called **net primary productivity** (**NPP**). All of these variables are measured in energy units (kilojoules) per square metre per year ($kJ\,m^{-2}\,year^{-1}$) fixed in photosynthesis or used in respiration.

The relationship between GPP, NPP and R is:

$$NPP = GPP - R$$

Energy transfer

Herbivores eat plants. The energy in the food is transferred from the primary producers (plants) to the herbivores. They use much of the energy in respiration for movement in the body. Some energy is lost as heat to the environment. The rest is available for other animals or decomposers. This can all be summarised in an energy flow diagram.

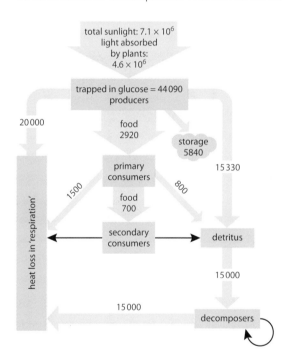

Energy flow through the trophic levels of a forest ecosystem. Units $kJ\,m^{-2}\,year^{-1}$ (kJ = kilojoules).

From this diagram we can calculate the efficiency of energy flow from one trophic level to the next, for example from producers to primary consumers. First we have to find out how much energy is available to primary consumers:

energy trapped (GPP)	−	energy plants use in respiration (R)	=	net primary productivity (NPP)
44 090	−	20 000	=	$24\,090\,kJ\,m^{-2}\,year^{-1}$

The transfer efficiency from producers to primary consumers is the amount transferred to the primary consumers, $2920\,kJ\,m^{-2}\,year^{-1}$ divided by the amount potentially available to them ($24\,090\,kJ\,m^{-2}\,year^{-1}$). So the answer is:

$$\text{efficiency of transfer} = \frac{2920}{24\,090} \times 100 = 12.12\%$$

Energy transfer efficiencies between trophic levels vary greatly in different ecosystems. A rule-of-thumb 10% is often quoted, but you will find values that differ widely from this.

Distribution and abundance

In any habitat a species occupies a specific **niche** determined by environmental conditions (**biotic** and **abiotic factors**) and the way that the species uses the habitat (food, shelter, sites, feeding times, etc.). The distribution (where they are) and abundance (how many) are determined by these conditions. Changes in these conditions can therefore lead to changes in distribution and abundance.

Succession

Primary succession happens when an area which is devoid of life is first **colonised** by species (usually lichen and algae on bare rock) that can cope in the harsh conditions. These are called **pioneer species**. They alter the environment in a way that makes it an unsuitable home for them, but suitable for new species to establish. The new species often replace the existing species. A similar process occurs time and again, through stages known as **seres**, until a stable community is reached. In stable woodland, for example, trees die but new ones of the same species grow to fill the gap. This is a **climax community**. If the succession starts with living things already present, for example if grazing stopped in a meadow, which then became woodland, this is called secondary succession.

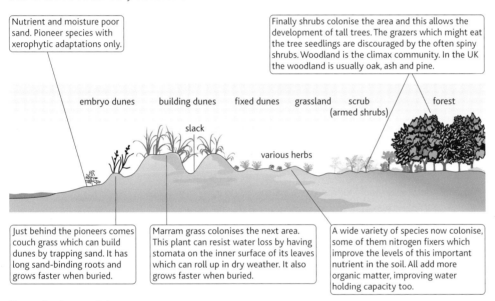

Nutrient and moisture poor sand. Pioneer species with xerophytic adaptations only.

Finally shrubs colonise the area and this allows the development of tall trees. The grazers which might eat the tree seedlings are discouraged by the often spiny shrubs. Woodland is the climax community. In the UK the woodland is usually oak, ash and pine.

embryo dunes   building dunes   fixed dunes   grassland   scrub (armed shrubs)   forest

slack

various herbs

Just behind the pioneers comes couch grass which can build dunes by trapping sand. It has long sand-binding roots and grows faster when buried.

Marram grass colonises the next area. This plant can resist water loss by having stomata on the inner surface of its leaves which can roll up in dry weather. It also grows faster when buried.

A wide variety of species now colonise, some of them nitrogen fixers which improve the levels of this important nutrient in the soil. All add more organic matter, improving water holding capacity too.

Succession in a sand dune system.

Quick Questions

Q1 Rewrite the equation NPP = GPP – R in terms of R, i.e. R = ?

Q2 What is a niche?

Thinking Task

Q1 Here is another example of part of an energy flow diagram:

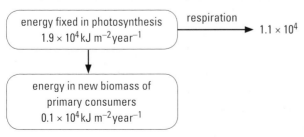

energy fixed in photosynthesis
1.9×10^4 kJ m^{-2}year^{-1}

respiration
→ 1.1×10^4

energy in new biomass of
primary consumers
0.1×10^4 kJ m^{-2}year^{-1}

Energy flow from producers to consumers in a grassland.

Calculate the efficiency of transfer from producers to consumers.

ResultsPlus
Watch out!

Biodiversity increases as succession progresses, but at the end it may go down again, as just a few species dominate the climax.

Investigating numbers and distribution

How can we practically investigate where organisms live (**distribution**) and how many there are (**abundance**)? The answer depends on what kind of habitat we are in and what we want to find out. A core practical is to carry out such an investigation.

If there appears to be a change across the area, a **transect** is the preferred method. If two areas appeared different and we wanted to compare them, we could take random samples within each area. In both cases we would use a piece of equipment called a **quadrat** to estimate abundance.

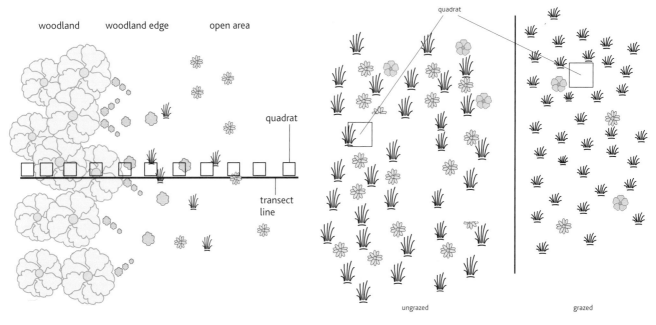

Left: A transect with quadrats used to investigate distribution and abundance of plant species on a woodland edge.
Right: Quadrats used to investigate distribution and abundance of plant species in grazed and ungrazed areas.

In either case, the usual methods to estimate abundance would be:
- *Either* count the individuals in a quadrat – this is not easily done with many plants, such as grasses, but quite possible with organisms such as limpets.
- *Or* find the percentage cover of each species – this is the most common method with plants. These estimates are best made using a quadrat that is divided up into smaller squares and counting the number of squares or part squares occupied by each species in turn. If, as is usual, there are 100 squares in the quadrat then the number of squares and part squares covered make up the percentage cover for that plant.

A different method involves the use of a point quadrat, in which pins are lowered systematically on to the vegetation, any 'hits' on the pins being recorded. These hits are added together to give percentage cover using the equation:

$$\% \text{ cover} = \frac{\text{hits}}{\text{hits} + \text{misses}} \times 100$$

In order to answer questions about the distribution and abundance patterns you have found in the habitat you are studying, you will also need to measure a number of factors:

Abiotic factor		Measurement technique
solar energy input		Use a light metre.
climate		Information about rainfall and temperature can be obtained from published sources.
topography		Topographical surveys measure the shape of the land. Surveyors' levelling equipment can be used – the simplest method being to use ranging poles and clinometers.
oxygen availability		Use an oxygen probe.
edaphic factors	pH	Use a pH probe or soil pH kit.
	minerals	Gardener's test kits can test the levels of important nutrients such as nitrate, phosphate and potassium (NPK).
	water	Soil sample can be weighed, dried slowly in an oven and reweighed to give the mass of water.
	organic matter	The dry soil sample can be weighed, burnt in a crucible and reweighed. Any organic matter is burnt off, which accounts for any difference in mass.
	soil texture	Soil texture charts can be used to assess if the soil is mainly clay, silt or sand.

Quick Questions

Q1 Explain what an edaphic factor is and give *three* examples.

Q2 Explain the meaning of these words:

niche transect quadrat

Q3 Choose a habitat you are familiar with and list *three* biotic and *three* abiotic factors which might control distribution and abundance of an organism in it.

Thinking Task

Q1 The table shows the data obtained in a study of a sand-dune system.

Site number	1	2	3	4	5	6
Distance from reference point/m	20	80	250	500	650	1800
Dead organic matter in soil/%	0.4	0.5	0.9	2.8	6.4	23.4
Number of plant species found	1	1	8	16	7	2

Plot the data in this table in a suitable form on one sheet of paper. Identify any patterns and comment on them.

Speciation and evolution

The theory of evolution is about how and why organisms have changed over time. What actually changes is **allele frequency** (the relative frequency of a particular allele in a population). New alleles arise from random changes in the DNA which makes up genes (**gene mutations**) and create variation within the population. Once a gene mutation has appeared it is acted upon by the selection pressures in the environment.

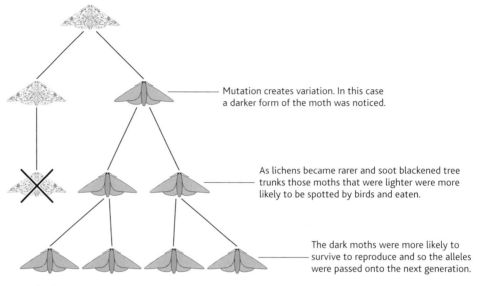

Mutation creates variation. In this case a darker form of the moth was noticed.

As lichens became rarer and soot blackened tree trunks those moths that were lighter were more likely to be spotted by birds and eaten.

The dark moths were more likely to survive to reproduce and so the alleles were passed onto the next generation.

Natural selection in peppered moths.

Speciation

If the ideas put forward above were all that was involved, species would change but there would be no new species. In order for a new species to form, part of an existing population must become **reproductively isolated** from another part. This usually happens when a barrier comes between two or more parts of an existing population. Over time, **natural selection** may cause the different parts of the population to change to such an extent that they can no longer interbreed to produce fertile offspring and this makes them two or more different species.

Prezygotic reproductive barriers	Explanation
habitat isolation	Populations occupy different habitats in the same area so do not meet to breed.
temporal isolation	Species exist in the same area but are active for reproduction at different times.
mechanical isolation	The reproductive organs no longer fit together.
behavioural isolation	Populations do not respond to each other's reproductive displays.
gametic isolation	Male and female gametes from two populations are simply incompatible with each other.
Postzygotic reproductive barriers	**Explanation**
hybrid sterility	Healthy individuals produced from the mating of two different species cannot themselves reproduce (e.g. the mule).
hybrid inviability	Individuals produced from the mating of two different species are not healthy and do not survive.

New evidence

Darwin's theory was very controversial in its day and still is for some people. There are now new types of evidence supporting the theory available to us:

- The DNA molecule is the same in all organisms. This supports Darwin's idea of descent from a common ancestor.
- DNA and proteins contain a record of genetic changes that have occurred by random mutations over time, indicating gradual change within and between species. By studying DNA (genomics) and proteins (**proteomics**) these changes can be identified. Comparing the DNA or amino acid sequences in different species can show how closely related species are in evolutionary terms. The more similar the sequence, the more closely related the species.
- Assessing the speed of mutation in DNA has shown that species have evolved over vast periods of time, as Darwin thought.

Validating evidence

Any new evidence must be carefully studied before it can be accepted. The scientific process has three key aspects which try to ensure reliability and validity:

- dedicated scientific journals
- peer review
- scientific conferences.

There are thousands of scientific journals published worldwide. Any research carried out must be published in at least one of these so that it can be read by other scientists. However, before it even gets to this stage it has to undergo a process called peer review. The editor of the journal sends a potential paper to two or three other scientists in the same area of work. They generally ask:

- Is the paper *valid*? (Are the conclusions based on good methods and are the data reliable?)
- Is the paper *significant*? (The paper must make a useful addition to the existing body of scientific knowledge.)
- Is the paper *original*? (Or has someone else already done the same work?)

Only if the other scientists agree that the paper is all these things can it be published. Conferences allow scientists to set out their ideas in front of other people who work in the same field. The suggestions can be assessed but there is no need to go through the peer review process.

Quick Questions

Q1 What must occur for speciation to take place?

Q2 Why is *survival* of the fittest not enough for evolution to happen?

Thinking Task

Q1 William Wilberforce was a fierce critic of Charles Darwin. He used the argument that no new species had ever been seen to arise. Even in dogs, where cross breeding had been prevented for many generations, the different breeds were still able to reproduce with each other, i.e. they were the same species.

Why does this *not* show that Darwin's theory is incorrect?

Greenhouse gases and the carbon cycle

There are a number of key questions to be asked:
- Are CO_2 levels rising and is there global warming?
- Does one cause the other?
- How bad will it get and can we do anything to combat it?

To answer these questions we need to look at evidence from many different sources.

What is the evidence for global warming?

Source of evidence	Link to global warming
temperature records	Long-term data sets allow changes in temperature to be analysed, e.g. the Central England Temperature series has records from 1659 to the present.
tree rings	Studying the size of tree rings is called **dendrochronology**. If the climate is warmer and wetter then the rings are wider. We can look at tree ring widths over 3000 years into the past and can tell a lot about the climate from them.
pollen data	Pollen grains are preserved in peat bogs. By sampling at different levels in the peat we are sampling at different ages. Analysis of the pollen can tell us which plants were growing and so what the climate was like when the peat was formed.
ice cores	Air trapped in ice when it was formed thousands of years ago can be analysed. This gives us information about temperatures and CO_2 levels in the past.

It seems that we are in a warm period of the Earth's history, but is this caused by rising CO_2 levels?

Carbon dioxide in the atmosphere is known as a greenhouse gas. It allows radiation to reach the Earth from the Sun. Some of this energy is trapped by CO_2 and the Earth warms up. This is the **greenhouse effect**. Other gases also have this effect, for example methane.

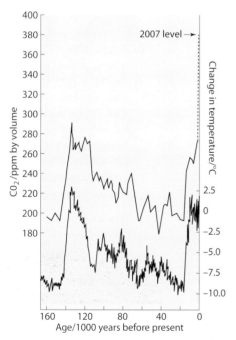

Changes in temperature and CO_2 concentrations over the past 160 000 years.

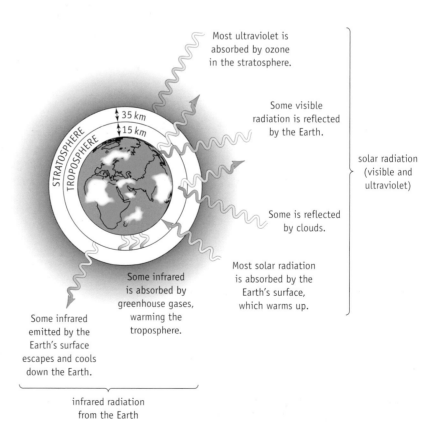

Inputs and outputs of energy to the Earth's atmosphere.

The data support the theory of global warming being caused by humans, due largely to CO_2 and methane emissions. So, the next question is, how bad will it get?

Computer modelling

Any attempt to predict climate change in the future must rely on very complex computer models to extrapolate from what we know to what might happen. These models get better all the time but they are limited by lack of computing power, sufficient data and knowledge of how the climate functions. Some factors such as carbon dioxide emissions or changes in ice cover are very hard to predict.

What can be done?

Science is telling us about a possible future problem, but it can also help to solve it. This is because we already have an understanding of the cycling of carbon in nature.

Plants take in CO_2 in photosynthesis, and trees store a lot of CO_2 as they gain size. Deforestation is thought to be an important cause of CO_2 increase in the air. The large-scale planting of new trees could reduce the amount of CO_2 in the atmosphere.

One way of reducing CO_2 levels would be to grow plants to use as fuel. They would only release the CO_2 they had just taken in and so would be carbon neutral. However, chopping down rainforest to grow palms for oil releases more CO_2 than it takes in, and using corn to make ethanol for biofuel deprives people of food.

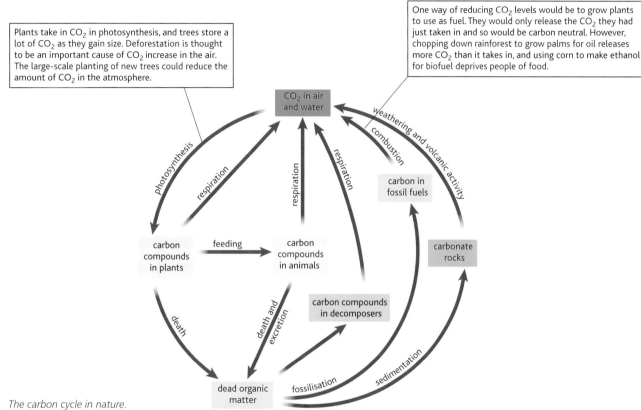

The carbon cycle in nature.

It is clear that there are a number of inputs and outputs to and from the atmospheric reservoir of CO_2. Until recently, the two have largely been in balance but now it is clear that there is extra CO_2 being added to the atmosphere by human activity. Looking at the cycle it can be seen that there are a number of ways we could intervene to offset this.

Reforestation will increase the removal of carbon dioxide from the atmosphere due to an increase in photosynthesis.

The increase in the use of biofuels as opposed to fossil fuels could also help because the CO_2 released in burning the fuel will only have been recently fixed in photosynthesis; their use is therefore carbon neutral.

ResultsPlus
Examiner tip

It is well worth learning the carbon cycle because you can work out such a lot from it.

Thinking Task

Q1 Sketch the carbon cycle and use it to explain how human activity has caused a build-up of carbon dioxide in the atmosphere.

Identify the processes that methods to reduce atmospheric carbon dioxide would influence.

Quick Questions

Q1 List *three* sources of evidence that can be used to investigate climate change.

Q2 What is the likely advantage, in terms of the greenhouse effect, of using biofuels? Suggest one disadvantage.

Impacts of global warming

Global warming will impact on the climate in many ways. In Britain a rise in temperature might mean warmer, wetter winters. Rainfall patterns are also likely to be effected. This might mean an increase in the risk of flooding in some areas. In other areas the risk might be drought. Finally, the seasonal cycles may change so that seasons are different in length and intensity. All of this will affect plants and animals in three main ways:

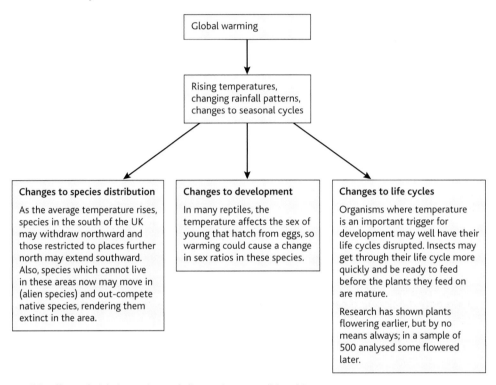

Possible effects of global warming and climate change on living things.

Temperature affects enzymes and therefore whole organisms – plants, animals and microorganisms.
Organisms may grow faster if temperatures rise by a few degrees. However, if the temperature rises too high their enzymes will denature and all reactions will stop.

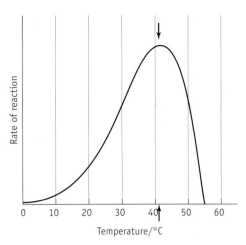

The effect of temperature on the rate of enzyme-catalysed reactions. The optimum temperature, which does not always have this value, is indicated by the arrows.

A practical investigation of the effect of temperature on development

We can investigate the effect of temperature on hatching rates in small invertebrates called brine shrimps. In this practical we want to vary temperature (*independent variable*) and see how it affects the number of shrimps hatched (*dependent variable*). There are other variables which might affect the rate such as salinity, pH and light level. It is important to keep these *controlled* or *monitored* during the experiment, both to make the data valid, and to care for the experimental animals. It is also possible to study the effect of temperature on seedling growth rates. In both experiments datalogging can be used to monitor the temperature and ensure reliable results.

Who decides?

There is little doubt that global warming is happening, but there are still big questions over what is causing it and what we should do about it. It is quite normal for scientists to disagree but this topic is also a matter for public debate. Non-scientists may not understand the uncertainty and naturally want a clear answer. The people who will give them this are often not the scientists but politicians, economists and other policy makers. Quickly the debate becomes politicised and then the usual impassionate methodology of science becomes sidelined.

It soon becomes clear that data are being interpreted with various hidden agendas and then this becomes the news rather than the science itself. So, scientists are accused of being funded by oil companies if they argue against the established political view, or politicised if they argue for it.

What conclusions people reach are often coloured by who funded the research they are doing, and pressures of economics and politics.

ResultsPlus
Watch out!

This enzyme temperature curve is not symmetrical; the effects of high temperatures are a sudden and complete denaturation of the enzyme, causing the curve to fall very steeply.

ResultsPlus
Examiner tip

Always consider the ethical aspects of experiments you do, especially those involving animals. There are often questions about this.

Quick Questions

Q1 What do we mean by an enzyme's optimum temperature?

Q2 Phenology is the study of seasonal events. Give *one* event that may be affected by rising temperature for animals and *one* for plants.

Q3 Why will increasing carbon dioxide levels not just be balanced by increased photosynthesis?

Thinking Task

Q1 List some of the ethical considerations when experimenting on hatching rates in brine shrimps.

Topic 5 – On the wild side checklist

By the end of this topic you should be able to:

Revision spread	Checkpoints	Spec. point	Revised		Practice exam questions
Photosynthesis 1	Describe chloroplast structure and how this is related to their job in photosynthesis.	LO2		☐	☐
	Describe photosynthesis as the splitting of water, storing of hydrogen in glucose and release of oxygen.	LO3		☐	☐
	Describe the light-dependent reactions that trap light energy by exciting electrons in chlorophyll, which are then involved in making ATP and reducing NADP.	LO4		☐	☐
Photosynthesis 2	Describe ATP manufacture from ADP using energy. ATP provides energy for biological processes.	LO5		☐	☐
	Describe the light-independent reactions in which carbon dioxide is reduced, using ATP and reduced NAD, and the manufacture of polysaccharides and other molecules.	LO6		☐	☐
Energy transfer abundance and ecosystems	Calculate net primary productivity from the equation: NPP = GPP – R.	LO7		☐	☐
	Calculate efficiency of energy transfer between trophic levels.	LO8		☐	☐
	Explain how biotic and abiotic factors control distribution and abundance of organisms.	LO10		☐	☐
	Explain how the concept of niche explains distribution and abundance of organisms.	LO12		☐	☐
	Describe succession to a climax community.	LO13		☐	☐
Investigating numbers and distribution	Describe an ecological study of a habitat to create valid, reliable data on abundance and distribution of organisms and measure a range of abiotic factors.	LO11		☐	☐
Speciation and evolution	Describe the role of gene mutation and natural selection in evolution; a change in allele frequency.	LO21		☐	☐
	Describe the validation of new evidence supporting the theory of evolution.	LO23		☐	☐
	Explain how reproductive isolation can lead to speciation.	LO22		☐	☐
Greenhouse gases and the carbon cycle	Outline possible causes of global warming.	LO14		☐	☐
	Discuss how understanding of the carbon cycle could help reduce atmospheric CO_2 levels.	LO9		☐	☐
	Analyse data from various sources to give evidence for global warming.	LO18		☐	☐
	Describe the modelling of trends in global warming and the limitations of this approach.	LO19		☐	☐
The effects of global warming	Describe the effects of global warming on plants and animals (distribution, development and life cycles).	LO15		☐	☐
	Explain the effects of increasing temperature on enzymes in living things.	LO16		☐	☐
	Describe an investigation of the effect of temperature on development of organisms; seedling growth or hatching in brine shrimps.	LO17		☐	☐
	Discuss how a person's point of view might affect the conclusions they reach about actions to take on global warming.	LO20		☐	☐

ResultsPlus
Build Better Answers

1 A transect can be used to study trends in the abundance and distribution of organisms.

Describe *one* method you could use to estimate the abundance of an organism at intervals along a transect line. (3)

☑ Examiner tip

This is a classic example of the need to read every word in the question very carefully.

A transect (*not* a quadrat) is used and you need to discuss *estimation* of distribution *and* abundance of an organism.

	Examiner comments
A line would be laid out and a quadrat placed along it. The number of species in each quadrat would then be counted and recorded down. The quadrat would then be randomly thrown and the number of species recorded down again until the finish.	This is quite a typical answer in which there is some truth, but certainly not enough to gain full marks. The use of a quadrat would gain a mark, although it is not the only device that could be used along the transect line. For example, if estimates of ground beetle numbers were being made, a pit-fall trap would be appropriate. The idea that the quadrat would be placed randomly, by throwing, is commonly held to be the way to use one. In this case random placing is not appropriate, the quadrat should be placed every half metre (if it is a half-metre by half-metre quadrat) – in a belt transect, or every so often (an interrupted belt transect). Finally, the thing to be estimated is not species number but the abundance of an organism, so the appropriate technique should be discussed for doing this, counting for discrete organisms like limpets or percentage cover for plants which are spreading.
	A **basic answer** would mention the quadrat but be unclear as to how to place or use it for a specific purpose.
	An **excellent answer** would describe laying out a line, placing quadrats along it either every so often or regularly and give full details of how to make accurate counts within the quadrat or make estimates of percentage cover within it.

2 Explain how some strains of bacteria have become able to survive treatment with antibiotics.

☑ Examiner tip

Note carefully the word 'become' in this question. Many answers discuss the ability of the bacteria to break down the antibiotic, pump it out of the cell, that they have a waxy coat or the ability to live inside host cells. Such an answer gains no credit at all.

Student answer	Examiner comments
The bacteria genetically mutate to gain resistance to the antibiotic. This might be because they now have a pump which pumps the antibiotic out of their cells. Some have an enzyme which breaks down the antibiotic. Some have a resistant cell wall.	This answer gets one mark for a reasonable statement about genetic mutation in bacteria which gives them resistance to antibiotics. The rest of the answer is not relevant to the question asked.
	A **basic answer** would mention genetic mutation as the origin of resistance or mention that antibiotics create a selection pressure when misused.
	An **excellent answer** would mention these points *and* go on to discuss how the selection pressure leads to antibiotic-resistant bacteria being able to reproduce, pass on the resistance allele and give rise to a population in which the frequency of this allele is increased.

Practice questions

1 The diagram below summarises the light-dependent reactions of photosynthesis.

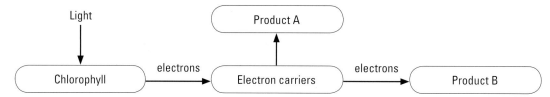

(a) Give the precise location within a chloroplast where this sequence of reactions occurs. (2)

(b) Give the names of product A and product B. (2)

(c) Give the name of the process that provides electrons to replace those lost by chlorophyll. (1)

(d) A chemical called atrazine prevents the flow of electrons to the electron carriers. Describe and explain the likely effect of atrazine on the production of carbohydrate in a chloroplast. (4)

Total 9 marks
(Biology (Salters-Nuffield) Advanced, June 2008)

2 *Agrostis tenuis* is a grass that grows near old copper mines in north Wales. Copper is usually very toxic to plants but some *Agrostis* plants can tolerate copper in the soil and grow on waste tips from the copper mines.

(a) Suggest a method for measuring copper tolerance in a sample of *Agrostis* plants. (2)

(b) Samples of tolerant and non-tolerant plants were grown in three trays of soil that contained no copper. Tray 1 contained only tolerant plants, Tray 3 contained only non-tolerant plants and Tray 2 had a mixture of equal numbers of both types. The total dry mass of the plants in each tray was measured. The arrangement of the plants and the results are summarised below.

	Tray 1 All tolerant plants	Tray 2 Mixed tolerant and non-tolerant plants	Tray 3 All non-tolerant plants
	× × × × × × × × × × × × × × × ×	× ○ × ○ ○ × ○ × × ○ × ○ ○ × ○ ×	○ ○ ○ ○ ○ ○ ○ ○ ○ ○ ○ ○ ○ ○ ○ ○
Total dry mass of tolerant plants	46 g	12 g	—
Total dry mass of non-tolerant plants	—	30 g	47 g

Suggest an explanation for the results obtained in tray 2. (4)

(c) Suggest and explain how the abundance of copper-tolerant *Agrostis* plants would be likely to change if the copper were removed from the soil on the mine waste tip. (2)

Total 8 marks
(Jan 2005, 6134/01 Question 6)

3 Sunflower seedlings were planted and kept under controlled conditions for 20 days. The gross primary productivity (GPP) and the net primary productivity (NPP) were measured each day. The results are shown in the graph below.

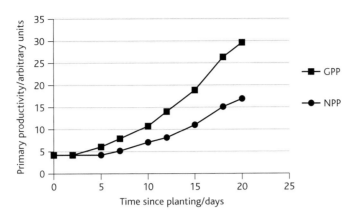

(a) (i) Compare the changes in GPP and NPP during the time period shown on the graph. (2)

(ii) Suggest an explanation for the changes you have described in (a)(i). (2)

(b) Explain the relationship between GPP, NPP and respiration. (2)

Total 6 marks
(Biology (Salters-Nuffield) Advanced, June 2007)

4 The tolerance of plants to copper ions in the soil is under genetic control. The frequency of an allele, which causes a plant to be more tolerant of copper, was measured at two different sites – A and B.

The table below shows the percentage frequencies of the tolerance and non-tolerance alleles in plant populations at the two sites.

(a) Explain what is meant by the frequency of an allele in a population. (2)

Site	Percentage frequencies of	
	Tolerance allele	**Non-tolerance allele**
A	30	70
B	80	20

(b) Describe how natural selection could have brought about the different allele frequencies at the two sites. (4)

(c) Suggest why bacteria often adapt to changing conditions much more quickly than plants. (2)

Total 8 marks
(Biology (Salters-Nuffield) Advanced, June 2007)

Decay and decomposition

Decay and decomposition, no matter how distasteful it might sometimes seem to us, is vital for the continuation of life on Earth. Plants need nutrients such as nitrogen, potassium, phosphorus and carbon to make biomass. These nutrients are locked into the tissues of the plants and any animals that might eat them. Once the plant or animal dies the nutrients can be released only through decay. The process of decomposition allows the nutrients to be recycled.

Micro-organisms are crucial to the decomposition process. The carbon cycle is a good example of how nutrients are recycled and how micro-organisms help. Bacteria and fungi produce a range of enzymes that are released on to the dead organic matter. The products of external digestion are absorbed by the micro-organism and broken down in microbial respiration, releasing carbon dioxide back into the atmosphere where it can be used again in photosynthesis.

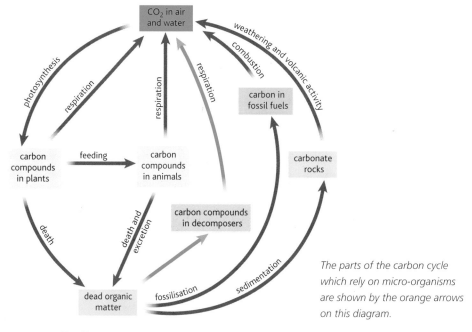

The parts of the carbon cycle which rely on micro-organisms are shown by the orange arrows on this diagram.

CSI Biology!

It is certainly possible to find out how long ago a mammal died. There are five main ways that scientists go about this.

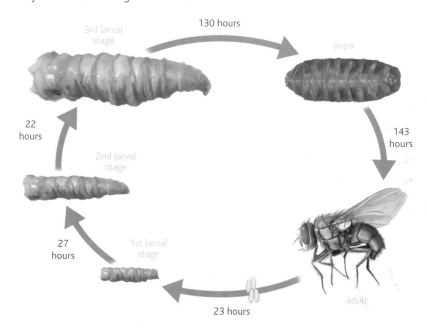

The time taken for the stages of the blowfly life cycle at 21 °C. The stage of larvae on a dead body can be used to estimate time of death.

Indicator of time of death	How a forensic scientist uses the information
body temperature	Body temperature is usually 37°C but the body begins to cool straight after death. During the first 24 hours after death the temperature of the body when it is found can be used to work out how long ago the person died.
degree of muscle contraction	After death, muscles usually totally relax and then stiffen. This stiffening is called rigor mortis. This happens within about 6–9 hours (depending on temperature). The stiffness occurs because muscle contraction relies on ATP, which cannot be made once respiration has stopped. So the muscles become fixed. The stiffness wears off again after about 36 hours in cooler conditions as the muscle tissue starts to break down.
extent of decomposition	Bodies usually follow a standard pattern of decay. Enzymes in the gut start to break down the wall of the gut and then the surrounding area. As cells die they release enzymes which help to break down tissues. The signs of decomposition, such as discoloration of the skin and gas formation, combined with information about environmental conditions allow time of death to be estimated.
forensic entomology	Determining the age of any insect maggots on the body allows the time the eggs were laid to be determined. This provides an estimate of time of death assuming any eggs were laid soon after death.
stage of succession	As a body decays, the populations of insects found on it change. There is a succession of species. The community of species present when the body is found allows the stage of succession to be determined and time of death estimated.

Putting all this information together can give the forensic scientist a very good estimate of time of death.

ResultsPlus
Examiner tip

All the methods used to indicate time of death of a body are subject to error. Remember this in examinations where you are asked to do calculations. Realistically, the answer will often be in the form of a range of possible times rather than a precise figure.

Quick Questions

Q1 List *three* indicators that can be used to work out time of death.

Q2 What role do micro-organisms play in the carbon cycle?

Thinking Task

Q1 Describe *two* ways in which succession on a body is similar to succession on a sand dune or other natural system and *one* way in which they are different.

DNA profiling

The DNA profiling (fingerprinting) technique was invented in 1985. Its influence on forensic science has been huge. The technique allows us to identify biological material with a high degree of confidence. In addition there is now also a technique called the polymerase chain reaction (PCR) which allows tiny amounts of DNA to be amplified into quantities large enough to use in DNA profiling. Together they form one of the most powerful techniques we have for criminal investigation and a range of other situations where total certainty about the identity of a sample is required.

Everyone's DNA is unique. This is because of the variety found in the sections of DNA which are not used to code for proteins. These non-coding sections are called **introns**. Scientists look for short, repeated sequences in these introns. The sequences of repeated bases are called **short tandem repeats** (STRs). There can be up to several hundred copies of the STR at a single locus. People (and all other organisms) vary in regard to the number of these repeats they carry at each locus. Scientists look at the short tandem repeats at many loci to build up a unique pattern for that individual.

DNA profiling procedure

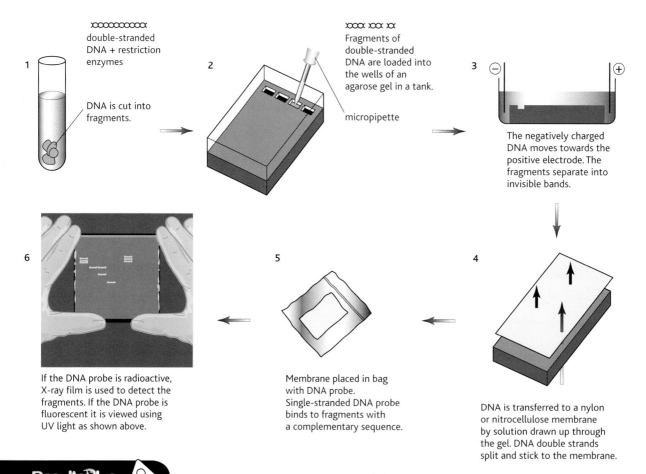

1 double-stranded DNA + restriction enzymes

DNA is cut into fragments.

2 Fragments of double-stranded DNA are loaded into the wells of an agarose gel in a tank.

micropipette

3 ⊖ ⊕
The negatively charged DNA moves towards the positive electrode. The fragments separate into invisible bands.

6 If the DNA probe is radioactive, X-ray film is used to detect the fragments. If the DNA probe is fluorescent it is viewed using UV light as shown above.

5 Membrane placed in bag with DNA probe. Single-stranded DNA probe binds to fragments with a complementary sequence.

4 DNA is transferred to a nylon or nitrocellulose membrane by solution drawn up through the gel. DNA double strands split and stick to the membrane.

The steps in the production of a DNA profile or fingerprint. The data can also be presented as a graph or as a series of numerical values. DNA fragments continuing the repeated sequences are created using reestriction enzymes or the polymerase chain reactions. DNA and DNA fragments carry a negative charge. If a potential difference is applied across a mixture of fragments in a suitable buffer, they will move towards the positive electrode. The fragments move at different rates according to their size and charge. Small fragments with fewer repeats travel faster and end up closer to the electrode after a set time. It is rather like chromatography, but here the separation is due to differences in size and charge, rather than solubility differences.

We have discussed DNA profiling in terms of forensic investigation to help the police to identify a criminal, but it has more uses than that. Confirming the pedigree of domestic animals (such as racehorses), looking at the purity of food samples (such as Basmati rice) and determining the father of a child, are all ways that the technique can be used. This is a good example of developments in science and technology allowing us to answer questions which we would previously not have been able to address.

PCR

The polymerase chain reaction (PCR) allows small samples of DNA to be amplified so that they can then be used in DNA profiling. The process relies on DNA primers, short sequences of DNA complementary to the DNA adjacent to the STR. A cycle of temperature changes results in huge numbers of the DNA fragments being produced.

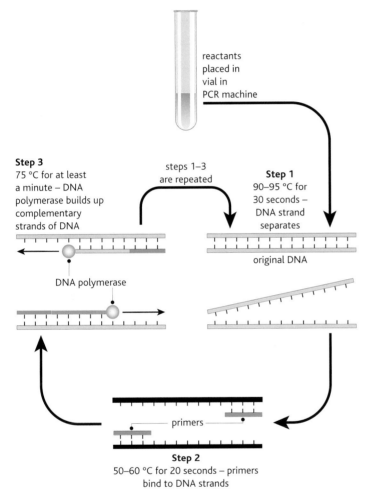

reactants placed in vial in PCR machine

Step 3
75 °C for at least a minute – DNA polymerase builds up complementary strands of DNA

steps 1–3 are repeated

Step 1
90–95 °C for 30 seconds – DNA strand separates

original DNA

DNA polymerase

primers

Step 2
50–60 °C for 20 seconds – primers bind to DNA strands

The procedure used for the polymerase chain reaction, PCR.

Quick Questions

Q1 Which parts of the DNA are used for profiling?

Q2 Why is it necessary for forensic scientists to look at 10 or more short tandem repeats when creating a DNA profile?

Thinking Task

Q1 The family tree of every racehorse in the UK can be traced back to just three male horses. Explain why it is easier to identify humans using DNA profiling than it is to identify racehorses.

DNA and protein synthesis

The basic steps of protein synthesis are as follows:

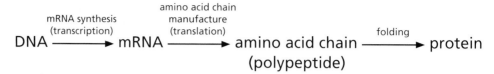

$$\text{DNA} \xrightarrow[\text{(transcription)}]{\text{mRNA synthesis}} \text{mRNA} \xrightarrow[\text{(translation)}]{\substack{\text{amino acid chain} \\ \text{manufacture}}} \substack{\text{amino acid chain} \\ \text{(polypeptide)}} \xrightarrow{\text{folding}} \text{protein}$$

Transcription

Transcription means 'to make a full copy of'. The mRNA copy is not direct, like a photocopy, but complementary. mRNA is made in the nucleus, but polypeptides are assembled in the cytoplasm, so the mRNA must move out of the nucleus through pores in the nuclear membrane.

Translation

Once in the cytoplasm, the mRNA attaches to a ribosome. tRNAs carrying specific amino acids bind to the complementary codons on the mRNA. A peptide bond is formed between neighbouring amino acids to produce a polypeptide. The whole process is shown below.

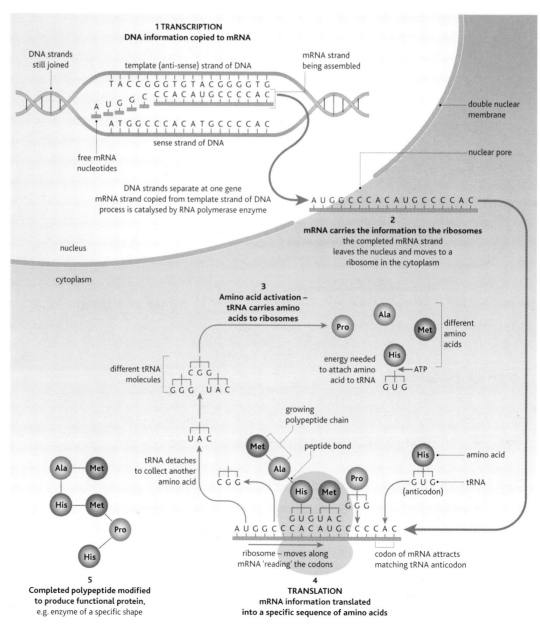

Protein synthesis is the way in which the information from a gene results in a sequence of amino acids.

The genetic code

The genetic code has several important properties:

Property of the genetic code	Why it is important
triplet code	With 20 amino acids and start and stop signals to code for and only four bases to do it, one base per amino acid will not do and neither will two. Using three bases gives 64 codons which is more than enough.
non-overlapping	Each set of three bases forms one triplet. The triplets do not overlap, so no base from one triplet is part of another triplet, avoiding confusion about which amino acid is being coded for.
degenerate	Some amino acids have more than one codon. For example, there are four different codons for the amino acid proline. As long as the codon starts with CC the amino acid proline will be put into the polypeptide. The code is said to be **degenerate**. This offers some protection against mutation.

ResultsPlus
Watch out!

The ribosome has two sites on its surface, one for a tRNA already bound to a growing polypeptide and the other for the tRNA carrying the next amino acid to be attached to the chain. Amino acids are added one at a time in a repeating process.

Post-transcriptional changes

It was originally thought that each gene coded for one protein. We now know this is not correct. Most genes code for many proteins and this is achieved by post-transcriptional changes in the mRNA. These changes are made to the mRNA before it is used in translation. We now think of the mRNA made in the nucleus as pre-mRNA.

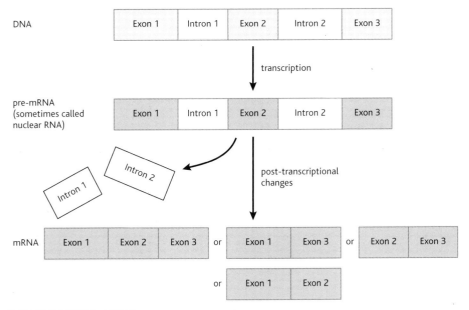

Different proteins can be made when different sections of mRNA are spliced together before translation.

Quick Questions

Q1 Which part of the DNA is used to code for proteins – intron or exon?

Q2 The DNA code is degenerate. Explain what this means and why it is significant.

Q3 Link each description in list A with the correct term in list B.

A	B
molecule with anticodon at one end and amino acid at the other	template strand
complementary copy of template strand	tRNA
DNA transcribed to mRNA	mRNA

Thinking Task

Q1 Draw a flowchart to explain the process of protein synthesis.

Infectious diseases and the immune response

Infectious diseases in humans are caused mainly by bacteria (singular: bacterium) and viruses. You need to be aware of the differences between them. Bacteria have a prokaryotic cellular structure with organelles but viruses do not.

Bacteria	Viruses
Cell surface membrane, cytoplasm, cell wall, ribosomes, plasmids and sometimes mesosomes, flagellum and pili.	No cell wall, cell surface membrane, cytoplasm or orgnelles. Nucleic acid enclosed in protein coat.
Circular strand of DNA is the genetic material.	DNA or RNA can be the genetic material.
Can live independently.	Must have a living organism as host.
Average diameter 0.5–5μm	20–400nm, wide range of sizes and shapes.
Often have mucus-based outer capsule.	May have outer membrane of host cell surface membrane, but containing glycoproteins from the virus.

Differences in structure between bacteria and viruses.

Tuberculosis and HIV

Both viruses and bacteria may enter the bodies of living things and, due to their own life processes, cause symptoms which can, in extreme cases, lead to death of the host organism. We can illustrate this by looking at tuberculosis, caused by the bacterium (*Mycobacterium tuberculosis*) and AIDS (caused by human immunodeficiency virus, HIV).

In TB the first infection may have no symptoms but tubercles form in the lungs due to the imflammatory response of the person's immune system. Some bacteria may survive inside the tubercles, due to their thick waxy coat. They lie dormant, but if the immune system is not working properly they can become active again. In active TB, lung tissue is slowly destroyed by the bacteria, causing breathing problems. The patient develops a serious cough, loses weight and appetite and may suffer from fever. TB bacteria also target cells of the immune system so the patient cannot fight other infections well. In some cases the bacteria invade glands and the CNS (central nervous system). All of this can be fatal.

The initial symptoms of an HIV infection are fevers, headaches, tiredness and swollen glands or there may be no symptoms. Three to 12 weeks after infection HIV antibodies appear in the blood and the patient is said to be HIV positive. All symptoms can then then disappear for years but eventually patients suffer from weight loss, fatigue, diarrhoea, night sweats and infections such as thrush. Finally there are severe symptoms such as dementia, cancers (e.g. Kaposi's sarcoma) and opportunistic infections such as TB and pneumonia.

Non-specific responses to infection

In order to fight a disease the body can react in a number of ways. The following responses are non-specific:

Response	inflammation	lysozyme action	interferon	phagocytosis
How it fights the infection	Damaged white cells release histamines that cause arterioles to dilate and capillaries to become more permeable. Blood flow to the area increases and plasma, white blood cells and antibodies leak out into tissues.	An enzyme found in tears, sweat and the nose destroys bacteria by breaking down the *bacterial* cell walls.	A chemical released from cells stops protein synthesis in *viruses*.	White blood cells engulf, digest and destroy *bacteria* and foreign material. These phagocytes include neutrophils and monocytes (which become macrophages).

Specific responses to infection

The specific immune response relies on the **lymphocytes** (another type of white cell), of which there are two main kinds, each with a number of sub-types. Both types respond to foreign (non-self) antigens such as proteins on the surface of bacteria and viruses. Macrophages are also involved, engulfing bacteria and displaying the non-self antigens. They alert the immune system to the presence of the foreign antigens. When any cell in the immune system displays antigens in this way, it is called an **antigen-presenting cell**.

The relationship between all these cell types and what they do is shown in the diagram.

ResultsPlus
Examiner tip

Never try to answer an exam question with just a diagram (unless it specifically asks for one). Use a diagram to help you to understand and then you can put it into words in the exam.

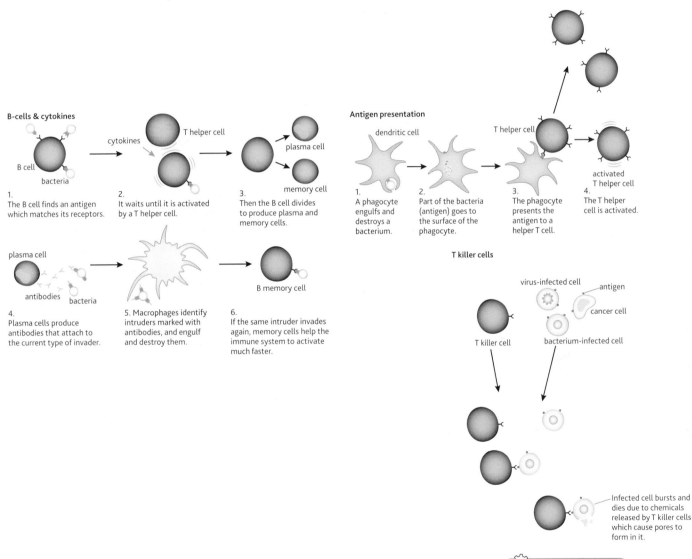

The immune response.

Quick Questions

Q1   Give *two* differences and *two* similarities between viruses and bacteria.

Q2   Why might a fever be a good thing if you have an infectious disease?

Thinking Task

Q1 Draw a table to distinguish between the roles of B cells (including B memory and B effector cells) and T cells (T helper, T killer and T memory cells).

Infection, prevention and control

Pathogens (organisms that cause diseases) enter the body through areas not covered by skin: nose, mouth, gas exchange surfaces, the eyes, gastrointestinal tract and genital tract. The entry of micro-organisms through wounds is also a major cause of infections.

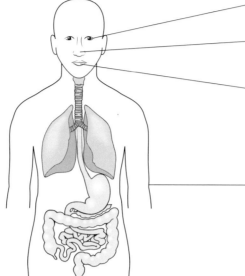

Eyes – tears contain the enzyme lysozyme which helps to digest microbes.

Respiratory tract – contains mucus which traps bacteria. The mucus is then swallowed and passed into the digestive system.

Gastrointestinal tract – acid in the stomach helps to protect against any microbes which are eaten. In addition the gut has its own bacteria. These compete with pathogens for food and space which helps to protect us. The harmless bacteria also excrete lactic acid which deters pathogens.

Skin – the skin is a tough barrier and usually only allows pathogens to enter if it is cut. As an additional line of defence the skin has its own microbes. These live naturally on the skin and out-compete pathogens. Sebum is an oily fluid which is made by the skin and can also kill microbes.

Major routes of entry of pathogens and the role of barriers in protecting the body from infection.

How can we develop immunity?

We can acquire immunity in different ways:

Type of immunity	Active		Passive	
	Natural	**Artificial**	**Natural**	**Artificial**
how it works	Exposed to antigen by getting the disease. The body produces memory cells which make it immune to disease in the future.	The injection of dead or weakened disease organisms, toxins or antigen fragments means that the body is exposed to the antigen and produces memory cells.	A mother's antibodies cross the placenta and are also found in breast milk. These antibodies can protect against any invading pathogen that the mother has encountered.	Injected with antibodies that can provide immediate protection against the invading pathogen they are specific for.

We can also use antibiotics to fight bacterial infections. There are two kinds of antibiotics: an antibiotic that kills bacteria is **bactericidal**, and one that limits or slows the growth of bacteria is **bacteriostatic**. When bacteria are no longer affected by an antibiotic they are said to be **resistant** to it.

Evolutionary race

As quickly as we evolve mechanisms to combat pathogens, they evolve new methods to overcome our immune system. The bacterium which causes TB and the virus responsible for AIDS (HIV) have both evolved features which help them to evade the immune system. The TB bacterium produces a thick waxy coat which protects it from the enzymes of the macrophages. The protein coat of HIV is constantly changing which means that the immune system can't target and destroy it.

Hospital acquired infections

Mutations also help some bacteria to become resistant to antibiotics which result in problems with antibiotic-resistant bacterial infections in hospitals. Hospitals try to combat this in a number of ways.

Hospital code of practice	How it helps to stop antibiotic resistance
only use antibiotics when needed and ensure course of treatment is completed	reduces selection pressure on organisms and destroys all bacteria causing infection
isolating patients with resistant diseases	prevents transmission of resistant bacteria between patients
good hygiene encouraged, including hand washing and bans on wearing of jewellery, ties and long sleeved shirts	prevents the spread of infection and cuts down on the number of places that may harbour pathogens
screening of patients coming into hospital	a person may be infected without showing symptoms; this can be detected and they can be isolated and treated

Investigating bacteria and antibiotics

A simple procedure can be carried out to investigate bacteria and antibiotics for the core practical. It is important to follow safe, aseptic techniques when doing this kind of work and essential to carry out a risk assessment.

> A *sterile* nutrient agar plate is seeded with suitable bacteria, e.g. using a *sterile* spreader.

↓

> Apply antibiotic to a *sterile* filter paper disc, then lay on the bacterial lawn using *sterile* forceps
> OR
> place antibiotic solution into a well in agar, using a *sterile* pipette.

↓

> *Seal Petri dish but do not tape all round the dish*

↓

> Incubate *below 30 °C* for about 24 hours.

↓

> Look for clear areas around the antibiotic discs or wells. Bigger areas indicate a better antibiotic against this bacterium species.

This flowchart shows how you investigate the effect of different antibiotics on bacteria. Anything in italics relates to aseptic procedure.

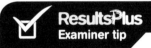
ResultsPlus
Examiner tip

Bacteria do not have a rapid rate of mutation, but they do have a high rate of reproduction. This means that new mutations occur quite regularly and each new mutation may help the organism to survive. Consequently they will produce many mutations in the time it would take a drug company to develop a new antibiotic.

ResultsPlus
Watch out!

When describing practical procedures you need to be very detailed. For example, if you say 'the bacteria are placed on a Petri dish' you will not get any marks. This is because bacteria will not grow in an empty plastic dish. You need to say 'nutrient agar' (in a Petri dish). Lack of details like this often lose candidates marks in exams.

Quick Questions

Q1 What theory would you hope to test with the practical above? Write a scientific question to answer.

Q2 Explain why the bacteria are incubated below 30 °C and why the dish is not completely sealed.

Thinking Task

Q1 Explain why it is difficult for drug developers to produce treatments that are effective in the long term.

Topic 6 – Infection checklist

By the end of this topic you should be able to:

Revision spread	Checkpoints	Spec. point	Revised	Practice exam questions
Decay and decomposition	Describe how to estimate time of death in a mammal using information from insect colonisation, muscle state and body temperature.	LO20	☐	☐
	Describe how micro-organisms are involved in decomposition.	LO9	☐	☐
DNA profiling	Describe the way in which DNA profiling (fingerprinting) can be used to work out genetic relationships in living things.	LO5	☐	☐
	Describe how the polymerase chain reaction can be used to give more DNA for analysis.	LO6	☐	☐
	Describe how DNA fragments can be separated by electrophoresis.	LO7	☐	☐
DNA and protein synthesis	Explain the process of protein synthesis and the role of DNA strands and RNA in it.	LO3	☐	☐
	Explain that the genetic code is triplet, non-overlapping and degenerate.	LO2	☐	☐
	Explain how changes that happen after transcription can give rise to more than one protein from one gene.	LO4	☐	☐
Infectious diseases and the immune response	Distinguish between the structure of bacteria and viruses.	LO8	☐	☐
	Explain the symptoms of bacterial and viral disease to include those of TB and infection by HIV.	LO11	☐	☐
	Describe non-specific immune responses: inflammation, action of lysozyme, interferon and phagocytosis.	LO12	☐	☐
	Explain antigens and antibodies and how plasma cells, macrophages and APCs are involved in the immune response.	LO13	☐	☐
	Distinguish between roles of T (helper, killer and memory) and B (memory and effector) cells.	LO14	☐	☐
	Explain how people develop natural, artificial, active and passive immunity.	LO15	☐	☐
Infection prevention and control	Describe the routes of entry of pathogens into the body and skin, stomach acid and gut/skin microbe barriers.	LO10	☐	☐
	Discuss the theory of an evolutionary race between pathogens and hosts and the evidence for this theory coming from HIV and TB.	LO16	☐	☐
	Distinguish between bacteriostatic and bactericidal antibiotics.	LO17	☐	☐
	Describe investigations of the effects of antibiotics on bacteria.	LO18	☐	☐
	Describe the way in which our understanding of infections acquired in hospitals can be used to reduce their occurrence through codes of practice.	LO19	☐	☐

ResultsPlus
Build Better Answers

1 Suggest why a patient infected with TB is more likely to develop symptoms of the disease if they are also infected by HIV. (2)

☑ Examiner tip

In order to gain full marks you are expected to write answers which contain detailed biological knowledge and relate different areas to each other.

	Examiner comments
HIV weakens the immune system and therefore it is likely that opportunistic infections such as TB would take advantage of a weakened immune system and therefore become more active. Symptoms will develop as the immune system response is suppressed and cannot efficiently destroy the bacterium, resulting in symptoms.	This is quite a typical answer which is not wrong but simply does not have enough detail to gain either of the two marks. A 'weakened immune system' does not address the specific aspect of the immune system which HIV affects. It is not TB which does anything in relation to the immune system, but the bacterium that causes it.
	A **basic answer** would mention that TB bacteria are not destroyed by the immune system.
	An **excellent answer** would link the destruction of T helper cells by HIV to the fact that the TB bacteria are not destroyed.

2 Give *two* symptoms which are likely to occur in a person with TB. (2)

☑ Examiner tip

On the face of it this is a very simple question, but again to gain marks at advanced level, answers must show some relevant detail.

Student answer	**Examiner comments**
Fever Coughing	This is an answer in which the student has thrown away a mark due to a simple lack of precision and careful thought. Almost everyone coughs at some point during most days so, to be a disease symptom, it should be obvious that this must be qualified.
	A **basic answer** would suggest a high temperature or fever as being a relevant symptom of TB.
	An **excellent answer** would suggest qualified coughing (i.e. coughing up blood or excessive coughing) as well as the appearance of tubercles, abnormal weight loss and development of fever.

Practice questions

1 The graph below shows the changes in population size of bacterial cultures grown in the presence of three antibiotics, A, B and C. In each case the antibiotic was added at 7 hours.

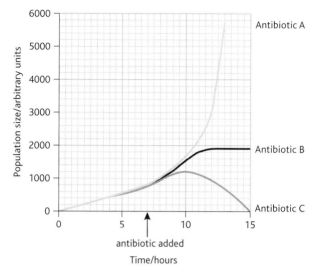

(a) Use examples from the graph to explain the differences between bactericidal and bacteriostatic antibiotics. (3)

(b) A previous investigation on the same bacterium using antibiotic A had produced a curve similar to that for antibiotic B. Suggest an explanation for the change in the response to antibiotic A. (4)

(c) Outline a technique that could demonstrate the effectiveness of antibiotics on bacteria. (4)

Total 11 marks
(Biology (Salters-Nuffield) Advanced, June 2008)

2 The hepatitis C virus (HCV) is transmitted in body fluids and infects the liver. HCV is very common in people who also have HIV infection. One treatment for HCV infection is injections of interferon.

(a) Explain why HCV infection is common in HIV positive people. (2)

(b) Name the type of cell involved in the normal immune response to virus-infected liver cells. (1)

Binding of interferon to infected cells causes an enzyme called PKR to become activated, and this prevents protein synthesis from occurring. The diagram below shows how interferon might be involved in the body's response to HCV infection.

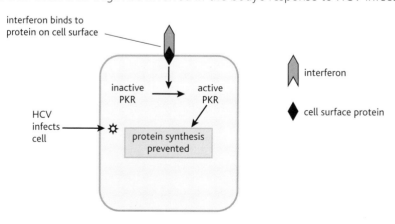

(c) With reference to the diagram above, explain the likely effects of interferon binding to the infected liver cell. (3)

Unfortunately, this treatment is only effective in 20% of cases because many strains of HCV are resistant to the effect of interferon. It has been found that these resistant viruses have a protein on their coats which inhibits the enzyme PKR.

(d) Suggest a reason why these virus strains are resistant to interferon. (2)

Total 8 marks

(Biology (Salters-Nuffield) Advanced, June 2006)

3 HIV can damage the human immune system.

   (a) Describe *two* active immune responses that are affected by HIV infection. (4)

   (b) Non-specific immune responses are not affected by HIV and can continue to prevent infection. Copy and complete the table below which shows some non-specific immune responses and descriptions of their functions.

Response	Description of function
inflammation	
	engulf and digest bacteria
lysosyme action	
	prevents viruses from multiplying

(4)

Total 8 marks

(Biology (Salters-Nuffield) Advanced, June 2006)

4 On 26th September, a forensic scientist was called to a room where a man was found dead. She was asked to determine the time of death.

She recorded the temperature in the room and she collected the larvae and pupae of several species of insect from the body. She took the pupae and larvae to her laboratory, where they were placed in a constant temperature of 23 °C.

On the 4th October, adults from four species of insect appeared, and another species appeared on the 6th October. One of the first species to be seen was the blowfly, which can lay eggs on a corpse within minutes of death, but which is rarely active at night. Records of weather conditions for the area were consulted and the time of death was determined to be 14th or 15th September.

   (a) Explain the importance of the temperature data in this investigation. (2)

   (b) Suggest one reason why collecting data about several species of insect would make the estimate of time of death more reliable. (1)

   (c) Suggest a reason why the scientist could not be more precise as to the time of death. (2)

Total 5 marks

(Biology (Salters-Nuffield) Advanced, June 2006)

Unit 4 specimen paper

1 The diagram below shows what happens to electrons during part of the light-dependent reactions of photosynthesis. Any excited electrons that are not taken up by electron carriers follow pathway A and release energy as light in a process called fluorescence. The excited electrons that are taken up by electron carriers follow pathway B.

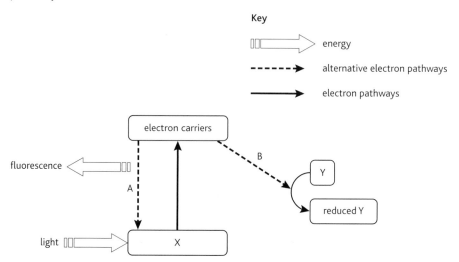

(a) Name the molecules X and Y shown on the diagram.   (2)

(b) Explain the importance of reduced Y in the process of photosynthesis.   (3)

(c) A light was shone on a leaf and left switched on.

The graph below shows changes in the amount of light given off as fluorescence by the leaf.

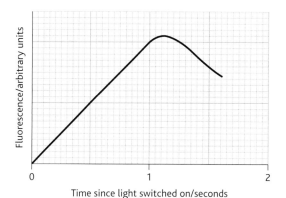

(i) Suggest an explanation for the increase in fluorescence.   (2)

(ii) Suggest a reason for the fall in fluorescence.   (1)

(d) Explain why an inhibitor of carbon dioxide fixation would lead to an increase in fluorescence.   (4)

Total 12 marks

(A2 6134 Biology Salters-Nuffield January 2009)

2 A study of tree pollen grains in a peat bog in Finland was carried out. The number of pollen grains of different tree species was recorded at different depths in the peat.

The data for four of these trees are given as a percentage of the total tree pollen sample, in the table below. An estimate of the age of the sample at each depth was also made.

Depth of sample/m	Age/years	Tree pollen grain/%			
		Larch	Spruce	Pine	Beech
0.5	2850	0	0	53	43
1.0	3770	0	0	55	40
1.5	5600	0	0	31	47
2.0	6390	0	12	15	53
2.5	8170	5	36	4	48
3.0	8700	38	36	6	35
3.5	8780	27	40	3	32
4.0	10000	10	22	2	40

The diagram below shows the present-day distribution of the four tree species found in the main climatic zones of the northern hemisphere.

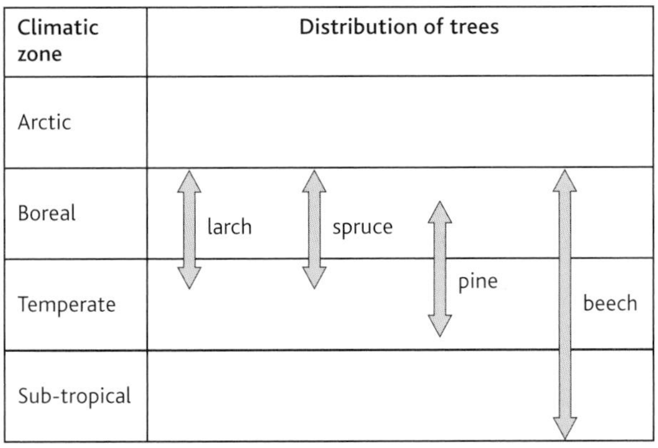

(a) Suggest how pollen grains can provide evidence about which species of tree were growing successfully in Finland as the peat bog was forming. (2)

(b) (i) Which species of tree listed below does not provide evidence about the changes in climate in Finland during the last 10 000 years? (1)

    A larch

    B spruce

    C pine

    D beech

    (ii) Explain your answer to (b) (i). (2)

(c) With reference to the present-day distribution of the four tree species and the results of the pollen grain study, suggest in what way the climate in Finland has changed during the last 10 000 years. Give reasons for your answer. (5)

(d) Describe how dendrochronology can be used to provide evidence for climate change. (2)

Total 12 marks

(A2 6BI04 Edexcel Specimen Paper)

3 Tuberculosis (TB) is caused by the bacterium, *Mycobacterium tuberculosis*.

(a) The table below lists five structural features that may be found in bacteria and viruses. Copy and complete the table by putting a cross in the box if the structural feature is present.

Structural feature	Bacteria	Viruses
mesosome	☐	☐
capsid	☐	☐
nucleic acid	☐	☐
cytoplasm	☐	☐
ribosome	☐	☐

(5)

(b) The table below shows the number of new TB cases recorded in 1994 and in 2004 from four different geographical regions. These data exclude people who are HIV positive.

Year	Number of new TB cases per 100 000 of the population			
	Africa	Asia	South America	Europe
1994	148	629	98	48
2004	281	535	59	104

(i) Describe the trends shown by the data. (2)

(ii) HIV positive people were excluded from the data. If they had been included suggest how the data would differ. Give an explanation for your answer. (3)

(c) TB is increasing in some countries which have well-funded health services. Suggest *two* reasons for this. (2)

Total 12 marks

(A2 6BI04 Edexcel Specimen Paper)

4 MRSA is a strain of the bacterium *Staphylococcus aureus*. MRSA can survive treatment with several antibiotics. An infection with MRSA is difficult to treat.

It is important to use an antibiotic that is effective against specific bacteria.

Describe in outline how you could test the effectiveness of an antibiotic on a specific bacterium in the laboratory. Include aspects of the method that ensure safe working. (5)

Total 5 marks

(A2 6134 Biology Salters-Nuffield January 2009)

5 An investigation was carried out to find the distribution of plant species on sand dunes. A transect was used which extended inland from a beach. Quadrats were used at nine positions along the transect. The percentage cover of selected species was recorded in each quadrat as well as the number of plant species in each quadrat. A sample of soil was taken from the area within each quadrat and used to measure the mass of organic material present.

The results are shown in the two tables below.

Quadrat number	1	2	3	4	5	6	7	8	9
Distance from top of beach/ metres	0	80	170	250	500	650	980	1600	1980
Number of species found	1	1	5	11	18	7	5	6	7
Mass of organic material/grams	0.4	0.3	0.3	0.9	2.8	6.4	25.1	23.4	32.8

	Percentage cover								
Quadrat number	1	2	3	4	5	6	7	8	9
Bare sand	80	30	30	8					
Sea couch	20								
Marram grass		70	50	20	5	5			
Red Fescue			5	40	55	40			
Sea buckthorn							80		
Common heather								90	
Corsica pine									100

(a) Explain why it is necessary to use a quadrat to estimate percentage cover. (2)

(b) Explain why a transect is more appropriate than random sampling in this study. (2)

(c) Use the information in both tables to compare the data collected from quadrat 1 and quadrat 5. (3)

(d) Differences in the variety and number of plant species found in the different quadrats can be explained by succession. Use the information in the table to suggest how the results of the study could be explained by succession. (5)

Total 12 marks

(A2 6134 Biology Salters-Nuffield Jan 2008)

Muscles and movement

Muscles, joints and movement

Bones can move in relation to one another at **joints**. Different types of joint allow different degrees of movement. **Ligaments** are made of elastic connective tissue. They hold bones together and restrict the amount of movement possible at a joint. **Tendons** are cords of non-elastic fibrous tissue that anchor muscles to bones.

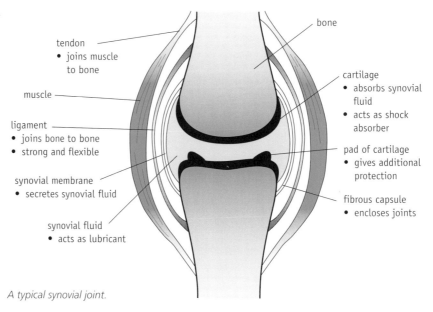

tendon
• joins muscle to bone

muscle

ligament
• joins bone to bone
• strong and flexible

synovial membrane
• secretes synovial fluid

synovial fluid
• acts as lubricant

bone

cartilage
• absorbs synovial fluid
• acts as shock absorber

pad of cartilage
• gives additional protection

fibrous capsule
• encloses joints

A typical synovial joint.

Skeletal muscles are those attached to bones and are normally arranged in **antagonistic pairs**. This means that there are pairs of muscles which pull in opposite directions. **Flexors** contract to flex, or bend a joint, e.g. biceps in the arm; **extensors** contract to extend, or straighten a joint, e.g. triceps in the arm.

Each skeletal muscle is a bundle of millions of muscle cells called fibres. Each muscle cell may be several centimetres long and contains several nuclei. It contains many **myofibrils** which are made up of the fibrous proteins **actin** (thin filaments) and **myosin** (thick filaments). The cell surface membrane of a muscle cell is known as the **sarcolemma**. The **sarcoplasmic reticulum** is a specialised endoplasmic reticulum which can store and release calcium ions. The cytoplasm inside a muscle cell is called the **sarcoplasm**. The specialised synapse (see page 63, Topic 8) between neurones and muscle cells is called the **neuromuscular junction**.

The sliding filament theory of muscle contraction

The functional unit of a muscle fibre is called a sarcomere. When the muscle contracts the thin actin filaments move between the thick myosin filaments, shortening the length of the sarcomere and therefore shortening the length of the muscle.

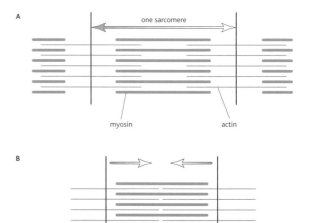

A

one sarcomere

myosin   actin

B

The arrangement of actin and myosin filaments in a sarcomere when relaxed (A) and contracted (B).

Myosin filaments have flexible 'heads' that can change their orientation, bind to actin and hydrolyse ATP (using **ATPase**). Actin filaments are associated with two other proteins, **troponin** and **tropomyosin**, that control the binding of the myosin heads to the actin filaments.

When a nerve impulse arrives at a neuromuscular junction, calcium ions are released from the sarcoplasmic reticulum and the following events take place that lead to the contraction of the muscle.

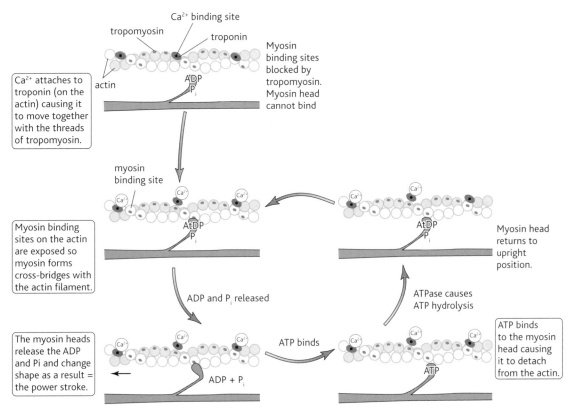

The sliding filament theory of muscle contraction.

Characteristics of fast-twitch and slow-twitch muscle fibres

Slow-twitch	Fast-twitch
specialised for slower, sustained contraction and can cope with long periods of exercise	specialised to produce rapid, intense contractions in short bursts
many mitochondria – ATP comes from aerobic respiration (electron transport chain)	few mitochondria – ATP comes from anaerobic respiration (glycolysis)
lots of myoglobin (dark red pigment) to store O_2 and lots of capillaries to supply O_2. This gives the muscle a dark colour	little myoglobin and few capillaries. The muscle has a light colour
fatigue resistant	fatigue quickly
low glycogen content	high glycogen content
low levels of creatine phosphate	high levels of creatine phosphate

Quick Questions

Q1 Give *one* reason why fast-twitch muscles are more likely to get tired faster than slow-twitch muscles.

Q2 Describe the role of ATP in muscle contraction.

Q3 Explain why muscles are arranged in antagonistic pairs.

Energy and the role of ATP in respiration

All living organisms have to respire. There are two different ways in which they do this – **aerobic respiration** (using oxygen) and **anaerobic respiration** (without oxygen). Both of these processes make the energy stored in glucose available in the form of **ATP**, to power **metabolic reactions**.

Aerobic respiration

$$\text{glucose} + \text{oxygen} \longrightarrow \text{carbon dioxide} + \text{water} + \text{energy}$$
$$C_6H_{12}O_6 + 6O_2 \longrightarrow 6CO_2 + 6H_2O + {\sim}30\ \text{ATP}$$

Anaerobic respiration

$$\text{glucose} \longrightarrow \text{lactic acid} + \text{energy}$$
$$C_6H_{12}O_6 \longrightarrow 2C_3H_6O_3 + 2\ \text{ATP}$$

The structure and function of ATP

ATP (adenosine triphosphate) is the cell's energy currency. Energy is required to add a third phosphate bond to ADP to create ATP. When this bond is broken by hydrolysis, the energy released can be used in energy-requiring processes taking place within the cell.

The breakdown of glucose in glycolysis

Starting with one glucose molecule, glycolysis produces two molecules of pyruvate, two molecules of reduced NAD and a net gain of two molecules of ATP. Glycolysis takes place within the cytoplasm of cells.

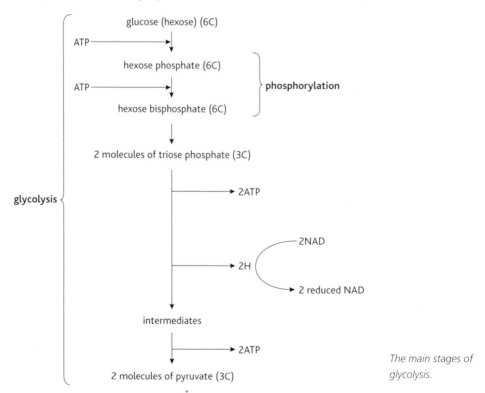

The main stages of glycolysis.

Anaerobic respiration

Glycolysis does not need molecular oxygen (O_2). However, for glycolysis to continue, a constant supply of NAD is required. In aerobic respiration the NAD is produced by the electron transport chain. The reduced NAD must be oxidised to NAD. During anaerobic respiration, NAD must come from elsewhere. In animals, pyruvate from glycolysis is reduced to give lactate, NAD is formed and can keep glycolysis going.

Anaerobic respiration allows animals to make a small amount of ATP. It is an inefficient process but it is rapid and can supply muscles with ATP when oxygen is not being delivered quickly enough to cells.

Lactate forms **lactic acid** in solution which lowers the pH. This can inhibit enzymes and, if allowed to build up, it can cause muscle cramp. Once aerobic respiration resumes most lactate is converted back to pyruvate. It is oxidised via the **Krebs cycle** into carbon dioxide and water. The extra oxygen required for this process is called the **oxygen debt**.

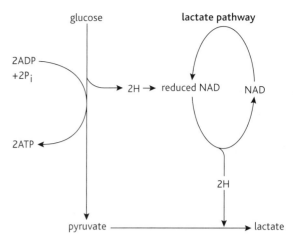

Anaerobic respiration in animals.

Investigating the rate of respiration using a respirometer

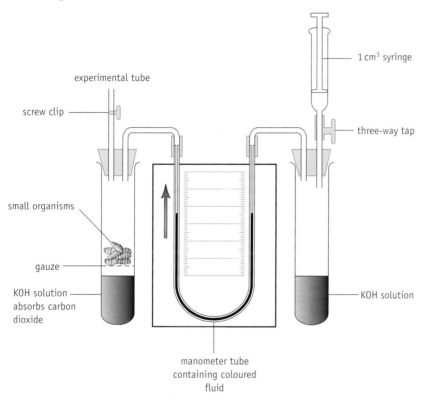

A respirometer

The rate of aerobic respiration can be determined using a respirometer by measuring the rate of oxygen absorbed by small organisms. Any CO_2 produced is absorbed by the potassium hydroxide (KOH) solution, so that any oxygen absorbed by the organisms results in the fluid in the manometer tube moving towards the organism (see arrow on diagram). The tube on the right-hand side compensates for any changes in pressure or temperature within the apparatus.

Quick Questions

Q1 Suggest *four* examples of biological processes that require the use of ATP.

Q2 Compare the role of ATP with glycogen.

Q3 Describe the role of NAD in anaerobic respiration.

**ResultsPlus
Examiner tip**

Don't forget the importance of including something to absorb the CO_2 or the respirometer reading will not change during aerobic respiration of carbohydrates because the same volume of gas is produced ($6CO_2$) as is absorbed by the organism ($6O_2$) per glucose molecule respired.

**ResultsPlus
Build Better Answers**

Remember that in the A2 Biology exams you may be asked to:
• bring together scientific knowledge and understanding from different areas
• apply knowledge and understanding of more than one area to a particular situation or context
• use knowledge and understanding of principles and concepts in planning experimental and investigative work and in the analysis and evaluation of data.

The respiration topic is a common choice for such synoptic questions because the process links to many other areas such as photosynthesis, food chains and muscle contraction.

Thinking Task

Q1 Draw the main stages of glycolysis alongside the main stages of the light-independent reactions of photosynthesis. Use these diagrams to identify the similarities and differences between the two processes.

Krebs cycle and the electron transport chain

Many of the reactions involved in respiration are redox reactions where one substrate is oxidised and another is reduced. When a molecule is oxidised, it either loses hydrogen or one or more electrons are lost. A molecule that gains electrons or hydrogen is reduced. One way of remembering this is to think of OILRIG (oxidation is loss, reduction is gain). When a molecule gains hydrogen it is reduced, and the molecule that loses the hydrogen is oxidised. For example: pyruvate → acetyl + 2H (is oxidation); NAD + 2H → reduced NAD (is reduction).

In aerobic respiration, the pyruvate (from glycolysis) is completely oxidised into carbon dioxide and water using oxygen.

Aerobic respiration takes place in two stages:

- First pyruvate is oxidised into carbon dioxide and hydrogen (accepted by the coenzymes **NAD** and **FAD**). This takes place in the **matrix** of the **mitochondria** and involves the **Krebs cycle**.

- In the second stage, most of the ATP generated in aerobic respiration is synthesised by **oxidative phosphorylation** associated with the **electron transport chain**. This involves **chemiosmosis** and the enzyme **ATPase**. It takes place on the **cristae** (inner membranes) of the mitochondria.

Preparation for the Krebs cycle (the link reaction)

In aerobic respiration each pyruvate molecule coming from glycolysis in the cell's cytoplasm enters the matrix of the mitochondrion. It is converted from pyruvate (3C) to an acetyl (2C) group. This involves the loss of CO_2 (decarboxylation) and hydrogen (dehydrogenation) generating reduced NAD. The acetyl group is carried by coenzyme A as acetyl coenzyme A.

The Krebs cycle

The Krebs cycle occurs in the matrix of the mitochondria. The main purpose of the cycle is to supply a continuous flow of hydrogen (and therefore electrons) to the electron transport chain for use in the synthesis of ATP by oxidative phosphorylation.

You do not need to know the names of the intermediate compounds of the Krebs cycle for the exam, but you are expected to appreciate that aerobic respiration is a many-stepped process with each step controlled and catalysed by a specific intracellular enzyme.

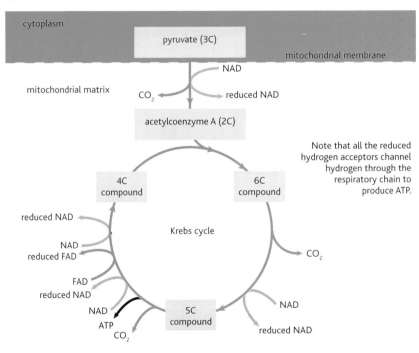

The reactions involved in the breakdown of pyruvate in aerobic respiration.

Each molecule of the 2-carbon acetyl coenzyme A from the link reaction is used to generate:

- three molecules of reduced NAD
- one molecule of reduced FAD
- two molecules of CO_2
- one molecule of ATP by **substrate-level phosphorylation** (synthesised directly from the energy released by reorganising chemical bonds)
- one molecule of a 4-carbon compound, which is regenerated to accept an acetyl group and start the cycle again.

Note that for each glucose molecule entering glycolysis two acetyl groups are formed, so the Krebs cycle will turn twice (i.e. producing two ATP and six reduced NAD, etc.)

Oxidative phosphorylation, chemiosmosis and the electron transport chain

Most of the ATP generated in aerobic respiration is synthesised by the electron transport chain.

1 Reduced coenzyme carries H+ and electron to electron transport chain on inner mitochondrial membrane.

2 Electrons pass from one electron carrier to the next in a series of redox reactions; the carrier is reduced when it receives the electrons and oxidised when it passes them on.

3 Protons (H+) move across the inner mitochondrial membrane creating high H+ concentrations in the intermembrane space.

4 H+ diffuse back into the mitochondrial matrix down the electrochemical gradient.

5 H+ diffusion allows ATPase to catalyse ATP synthesis.

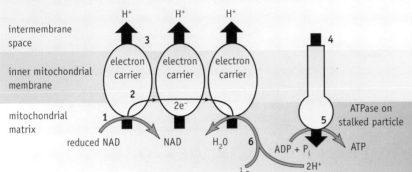

6 Electrons and H+ ions recombine to form hydrogen atoms which then combine with oxygen to create water. If the supply of oxygen stops, the electron transport chain and ATP synthesis also stop.

The electron transport chain and chemiosmosis result in ATP synthesis by oxidative phosphorylation.

The majority of ATP generated by aerobic respiration comes from the activity of the electron transport chain in the inner membrane of the mitochondria (cristae).

The overall reaction of aerobic respiration can be summarised as the splitting and oxidation of a respiratory substrate (e.g. glucose) to release carbon dioxide as a waste product, followed by the reuniting of hydrogen with oxygen to release a large amount of energy in the form of ATP.

Quick Questions

Q1 Describe what happens to the carbon and hydrogen atoms from a glucose molecule in aerobic respiration.

Q2 Explain what oxidative phosphorylation means.

Q3 Explain why the electron transport chain and the Krebs cycle would stop if there was no oxygen.

Thinking Task

Q1 Sketch a simple diagram of a cell and mitochondria and outline where the main steps in aerobic respiration take place.

The heart, energy and exercise

The control of the cardiac cycle

The impulse to contract originates within the heart itself from the sinoatrial node – the heart is said to be **myogenic**.

1 Electrical impulses from the SAN spread across the atria walls, causing contraction. This is called atrial systole.

2 Impulses pass to the ventricles via the AVN after a short delay to allow time for the atria to finish contracting.

3 Impulses pass down the Purkyne fibres to the heart apex.

4 The impulses spread up through the ventricle walls causing contraction from the apex upwards. Blood is squeezed into the arteries. This is ventricular systole.

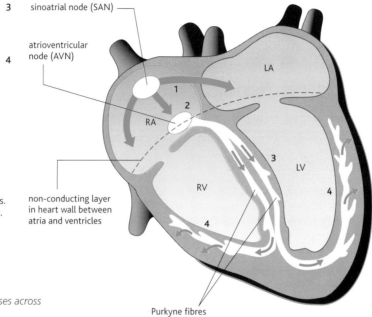

The route taken by electrical impulses across the heart during the cardiac cycle.

After contracting (**systole**), the cardiac muscle then relaxes for a period called **diastole** when the blood returning from the veins fills the atria.

Measuring electrical changes in the heart

Electrical currents caused by the spread of the electrical impulse (wave of depolarisation) during the cardiac cycle can be detected with an electrocardiogram (ECG).

If disease disrupts the heart's normal conduction pathways changes will occur in the ECG pattern which can be used for diagnosis of **cardiovascular disease**.

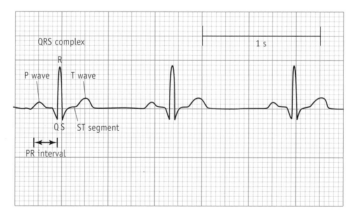

A normal ECG pattern in a healthy heart.

- The **P** wave is the time of atrial systole.
- The **QRS** complex is the time of ventricular systole.
- The **T** wave is caused by repolarisation of the ventricles during diastole.

Regulation of cardiac output

Blood is pumped around the body to supply O_2 and remove CO_2 from respiring tissues. How much is pumped in a minute (cardiac output) depends on two factors: how quickly the heart is beating (**heart rate**) and the volume of blood leaving the left ventricle with each beat (**stroke volume**).

cardiac output $(dm^3 min^{-1})$ = stroke volume (dm^3) × heart rate (min^{-1})

The heart rate can be affected by hormones (e.g. adrenaline) and nervous control. The **cardiovascular control centre** in the **medulla** of the brain controls the sinoatrial node via nerves. The **sympathetic nerve** speeds up the heart rate in response to falls in pH in the blood due to CO_2 and lactate levels rising, increases in temperature and mechanical activity in joints.

Impulses carried by the **vagus nerve** (parasympathetic) slow down the heart rate when the demand for O_2 and removal of CO_2 reduces.

Regulation of ventilation rate

The rate at which someone breathes is called the **ventilation rate**. This is often expressed as the volume of air breathed per minute (the minute ventilation). The volume of air breathed in or out of the lungs per breath is called the **tidal volume**. The maximum volume of air that can be forcibly exhaled after a maximal intake of air is called the **vital capacity**.

> ventilation rate = tidal volume × number of breaths per minute

The **ventilation centre** in the **medulla** controls the rate and depth of breathing in response to impulses from chemoreceptors in the medulla and arteries which detect the pH and concentration of CO_2 in the blood. Impulses are sent from the ventilation centre to stimulate the muscles involved in breathing. A small increase in CO_2 concentration causes a large increase in ventilation. It also increases in response to impulses from the motor cortex and from stretch receptors in tendons and muscles involved in movement. We also have voluntary control over breathing.

Measuring lung volumes using a spirometer

A person using a **spirometer** breathes in and out of an airtight chamber causing it to move up and down and leaving a trace on a revolving drum (kymograph).

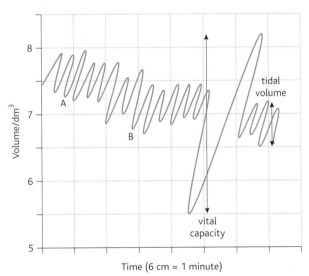

A spirometer trace showing quiet breathing with one maximal breath in and out.

SAFETY!
A canister of soda lime can be used to remove the CO_2 from the exhaled air to measure the volume of O_2 absorbed by the person after exercise, but it is important to remember that the chamber must be filled with medical grade O_2 before starting if this is to be attempted.

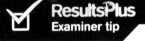

ResultsPlus
Examiner tip

You can calculate the volume of O_2 absorbed in a given time by measuring the differences in volume between the troughs labelled A and B in the diagram and dividing by the time between A and B.

Quick Questions

Q1 If someone takes 11 breaths per minute with an average tidal volume of $0.45\,dm^3$ calculate their ventilation rate.

Q2 Sketch what you would expect an ECG trace to look like if a patient suffered from ventricular fibrillation. (This is rapid and uncontrolled contractions in the ventricles sometimes caused by a patch of damaged tissue in the ventricle acting as a pacemaker.)

Q3 Suggest what might happen to the heart rate if the connection between the sinoatrial node and the vagus and sympathetic nerves was cut.

Thinking Task

Q1 Summarise and explain the effects of exercise on both heart and ventilation rate.

Homeostasis

Homeostasis is the maintenance of a stable internal environment, within a narrow limit, of the optimum conditions needed by cells so they can function properly. A homeostatic system therefore requires:

- **receptors** to detect the change away from the norm value (**stimulus**)
- a **control mechanism** that can respond to the information. The control mechanism uses the **nervous system** or **hormones** to switch **effectors** on or off
- **effectors** to bring about the response (usually to counteract the effect of the initial change). Muscles and glands are effectors.

input ⟶ receptors ⟶ control mechanism ⟶ effectors ⟶ output
feedback ◄

Negative feedback helps to keep the internal environment constant. A change in the internal environment will trigger a response that counteracts the change, e.g. a rise in temperature causes a response that will lower body temperature. For negative feedback to occur, there must be a **norm value** or **set point**, e.g. 37.5 °C for core body temperature.

A Conditions controlled by homeostasis fluctuate around the norm value.

norm value
(set point)

B The condition is controlled by negative feedback.

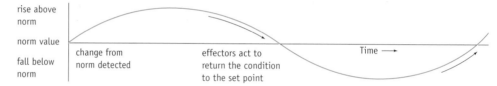

rise above norm

norm value

fall below norm

change from norm detected

effectors act to return the condition to the set point

Time ⟶

A summary of the role of negative feedback in maintaining body systems within narrow limits.

Homeostasis and exercise

We have already seen that the body responds to the demands of exercise by increasing cardiac output and ventilation rate under the control of centres in the medulla (see page 51 – The heart, energy and exercise). Not only does the increased respiration rate during exercise produce a lot of CO_2 and/or lactate, but the energy transfers also release a lot of heat energy. This can be enough for a 1 °C rise in body temperature every 5–10 minutes if we can't disperse the heat away from the body.

The control of core body temperature through negative feedback is called thermoregulation. Thermoreceptors in the skin detect changes in temperature. In addition thermoreceptors in the hypothalamus (in the brain) can detect changes in the core blood temperature. If a rise in temperature is detected above the norm value the heat loss centre in the hypothalamus will stimulate effectors to increase heat loss from the body – usually through the skin.

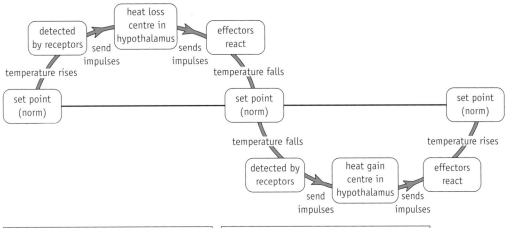

Negative feedback in thermoregulation.

Heat loss centre	
Stimulates:	• sweat glands to secrete sweat.
Inhibits:	• contraction of arterioles in skin (dilates capillaries in skin)
	• hair erector muscles (relax – hairs lie flat)
	• liver (reduces metabolic rate)
	• skeletal muscles (relax – no shivering).

Heat gain centre	
Stimulates:	• arterioles in the skin to constrict
	• hair erector muscles to contract
	• liver to raise metabolic rate
	• skeletal muscles to contract in shivering.
Inhibits:	• sweat glands.

Above or below certain temperatures homeostasis fails (e.g. because the hypothalamus may be damaged). Instead, positive feedback may occur resulting in a high temperature continuing to rise or a low temperature falling still further. This can result in hyperthermia or hypothermia and may lead to death.

Medical technology to enable those with injuries and disabilities to participate in sport

The development of **keyhole surgery** using fibre optics has made it possible for surgeons to repair damaged joints (including torn cruciate ligaments in the knee) with precision and minimal damage. This is because only a small incision (cut) is needed so there is less bleeding and damage to the joint, and recovery is much quicker.

A **prosthesis** is an artificial body part designed to regain some degree of normal function or appearance. The design of prostheses has improved significantly and many disabled athletes are now able to compete at a very high level, e.g. with dynamic response feet that can literally provide them with a spring in their step. Damaged joints (such as knee joints) can also now be repaired with small prosthetic implants to replace the damaged ends of bones, freeing the patient from a life of pain and restoring full mobility.

? Quick Questions

Q1 Explain what is meant by the term 'negative feedback'.

Q2 Suggest what the consequences might be if you were unable to lose heat from the body during exercise.

Q3 Describe the body's likely responses to the core temperature dropping below 37 °C.

⚙ Thinking Task

Q1 Using your revision in this section and pages 45, 50 and 51 explain why some animals are adapted to short bursts of fast or powerful exercise, while others are adapted to long periods of continuous exercise.

Health, exercise and sport

The possible effects of too little exercise

There are many benefits to regular moderate exercise. Here are a few possible effects of a lack of exercise over a prolonged period of time:

- reduced physical endurance, lung capacity, stroke volume and maximum heart rate
- increased resting heart rate, blood pressure and storage of fat in the body
- increased risk of **coronary heart disease**, type II diabetes, some cancers, weight gain and **obesity**
- impaired immune response due to lack of natural killer cells
- increased levels of LDL ('bad' cholesterol) and reduced levels of HDL ('good' cholesterol)
- reduced bone density, therefore increased risk of osteoporosis.

The possible effects of too much exercise

Overtraining can lead to symptoms such as immune suppression and increased wear and tear on joints. It can also result in chronic fatigue and poor athletic performance.

Too much exercise generally may also increase the amount of wear and tear on joints. Damage to **cartilage** in synovial joints can lead to inflammation and a form of arthritis. Ligaments can also be damaged. Bursae (fluid-filled sacs) that cushion parts of the joint can become inflamed and tender.

There is also some evidence of a **correlation** between intense exercise and the risk of infection such as colds and sore throats. This could be **caused** by an increased exposure to pathogens, or a **suppression of the immune system**. There is some evidence that the number and activity of some cells of the immune system may be decreased while the body recovers after vigorous exercise. It may also be the case that damage to muscles during exercise and the release of hormones such as adrenaline may cause an inflammatory response which could also suppress the immune system.

Some ethical positions relating to the use of performance-enhancing substances by athletes

Some athletes will do anything they can, in the pursuit of excellence. This might include the use of illegal performance-enhancing substances. Others may feel they need to follow suit because they don't want to be at a disadvantage. This has been a subject for debate in the sporting world for many years.

These ethical frameworks can be used when considering both sides of the argument:

- rights and duties
- maximising the amount of good in the world
- making decisions for yourself
- leading a virtuous life.

For example, doping in sport could be considered *not* acceptable because athletes have a right to fair competition, but could equally be considered acceptable because athletes have the right to exercise autonomy, for example to choose to achieve their best performance, and also have a duty to any sponsor they may have.

Remember that in order to maintain that something is ethically acceptable or not, you must provide a reasonable explanation as to why that is the case.

Ethical **absolutists** see things as very clear cut. They would take one of two positions:

1. It is never acceptable for athletes to use performance-enhancing substances (even if they are legal), *or*
2. it is always acceptable for athletes to use any substance available to them to compete more effectively, even if there are associated risks to their health.

Ethical **relativists** would realise that people and circumstances may be different, e.g:

- It is wrong for athletes to use performance-enhancing substances, but there may be some cases and circumstances where it is acceptable.

ResultsPlus
Examiner tip

Remember that exam questions in this unit may refer back to any other topics from the A level biology course, so now would be a good time to check your notes about the cells involved in the specific and non-specific immune system (page 33, Unit 4).

ResultsPlus
Watch out!

Just because two things are observed to happen, it doesn't mean they are connected. In particular it doesn't mean that one caused the other. A correlation does not necessarily mean a connection. If there appears to be a strong correlation between two factors, a causal link is more likely if you can provide a biological explanation for why one factor will affect the other, especially if there aren't many other likely factors or explanations available. For example, there is a positive correlation between the number of shark attacks and the number of ice creams sold at a beach. There is no biological explanation for this correlation, so there is no direct causal link. In contrast it is thought that there is a causal link between the number of cigarettes smoked and the number of deaths due to lung cancer, because there is a strong correlation and a biological explanation about why smoking could cause lung cancer.

How can drugs affect your genes?

Some drugs such as anabolic steroids are closely related to natural steroid hormones. They can pass directly through cell membranes and be carried into the nucleus bound to a receptor molecule. These hormone/receptor complexes act as **transcription factors**. They bind to the **promoter** region of a gene allowing **RNA polymerase** to start **transcription**. As a result more **protein synthesis** takes place in the cells. For example testosterone increases protein synthesis in muscle cells, increasing the size and strength of the muscle tissue. Peptide hormones do not enter cells directly, but they bind with receptors on the cell surface membrane. This starts a process that results in the activation of a transcription factor within the nucleus. For example erythropoietin (EPO) stimulates the production of red blood cells. This means that the blood can carry more oxygen which is an advantage for an athlete.

Genes are switched on by successful formation and attachment of the transcription initiation complex to the promoter region.

Genes remain switched off by failure of the transcription initiation complex to form and attach to the promoter region. This is due to the absence of protein transcription factor(s) or the action of repressor molecules.

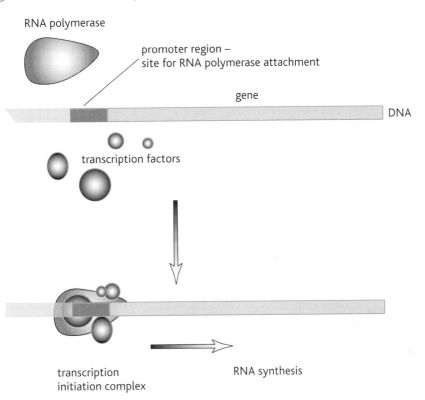

DNA transcription is controlled by transcription factors.

Quick Questions

Q1 Describe why a lack of exercise may lead to an increased risk of coronary heart disease.

Q2 Explain why a lack of T helper cells may increase the risk of an athlete suffering from a sore throat.

Q3 Outline the role of transcription factors in the control of gene expression.

Thinking Task

Q1 Even if all performance-enhancing substances were formally banned, would we ever have a level playing field for athletes?

Topic 7 – Run for your life checklist

By the end of this topic you should be able to:

Revision spread	Checkpoints	Spec. point	Revised		Practice exam questions
Muscles and movement	Describe the structure of a muscle fibre and explain the differences between fast and slow twitch muscle fibres.	LO2		☐	☐
	Explain how skeletal muscle contracts using the sliding filament theory.	LO3		☐	☐
	Recall the way in which muscles, tendons, the skeleton and ligaments interact to allow movement.	LO4		☐	☐
Energy and the role of ATP	Describe aerobic respiration as splitting of glucose to release carbon dioxide, water and energy.	LO5		☐	☐
	Describe a practical to investigate rate of respiration.	LO6		☐	☐
	Recall what ATP is and how it supplies energy for cells.	LO7		☐	☐
	Describe the roles of glycolysis in both aerobic and anaerobic respiration. You do not need to know all the stages but you do need to know that glucose is phosphorylated and ATP, reduced NAD and pyruvate are produced.	LO8		☐	☐
	Explain what happens to lactate after you stop exercising.	LO11		☐	☐
The Krebs cycle and the electron transport chain	Describe how the Krebs cycle produces carbon dioxide, ATP, reduced NAD and reduced FAD. You should also understand that respiration has lots of enzyme-controlled steps.	LO9		☐	☐
	Describe how ATP is made by oxidative phosphorylation in the electron transport chain including the roles of chemiosmosis and ATPase.	LO10		☐	☐
The heart, energy and exercise	Understand that cardiac muscle is myogenic and describe how electrical activity in the heart allows it to beat. You should also know how ECGs can be used.	LO12		☐	☐
	Explain that tissues need rapid delivery of oxygen and removal of carbon dioxide during exercise and that changes in ventilation and cardiac output allow this to happen. You should understand how heart rate and ventilation rate are controlled.	LO13		☐	☐
	Describe how to use data from spirometer traces to investigate the effects of exercise.	LO14		☐	☐
Homeostasis	Explain the principle of negative feedback.	LO15		☐	☐
	Discuss the concept of homeostasis and how it maintains the body during exercise, including controlling body temperature.	LO16		☐	☐
Health, exercise and sport	Explain how genes can be switched on and off by DNA transcription factors including hormones.	LO17		☐	☐
	Analyse and interpret data on the possible dangers of exercising too little and too much. You should also be able to talk about correlation and cause.	LO18		☐	☐
	Explain how medical technology helps people with injuries or disabilities to take part in sport.	LO19		☐	☐
	Outline the ethics of using performance-enhancing substances.	LO20		☐	☐

ResultsPlus
Build Better Answers

1 Animals that are predators often show bursts of very fast movement. Their prey tend to be able to carry out sustained movement over longer periods of time. Close examination shows that the muscles of predator and prey show a different composition of fast- and slow-twitch fibres.

  (a) (i) Outline the differences between fast- and slow-twitch muscle fibres. (2)

  (ii) State whether predator or prey would show a higher proportion of slow-twitch fibres. (1)

  (iii) Discuss why predators show different proportions of fast- and slow-twitch muscle fibres from their prey. (2)

☑ **Examiner tip**

If you are asked for the differences, make sure you refer to both or use a comparative word, e.g. 'more'.

Student answer	Examiner comments
(a) (i) Slow-twitch muscle fibres have more mitochondria and more capillaries supplying oxygen than fast twitch fibres. (ii) Prey. (iii) Predators are likely to have more fast-twitch than slow-twitch fibres, in comparison to their prey. This is because predators tend to be fast and powerful over short distances to catch and kill their prey and therefore use anaerobic respiration to release ATP quickly.	This is a good response because not only does it provide a likely comparison, it also provides a clear and plausible explanation.

(b) During fast movement, lactate builds up in the muscles of a predator, such as a cheetah. Explain what happens to this lactate after the chase has ended. (3)

Student answer	Examiner comments
Lactate diffuses from the muscle into the blood where it is carried away from the muscle to prevent cramp.	This response is a correct but only partial explanation. It explains how the lactate is moved away from the muscle, but not how it is removed from the body.
Lactate is oxidised back into pyruvate using NAD that has been oxidised in the electron transport chain using oxygen. The extra oxygen needed is the oxygen debt.	This response will gain maximum marks because it provides a chemical explanation of the fate of the lactate, clearly demonstrating an understanding of both aerobic and anaerobic respiration, as well as recognition of the need for extra oxygen.

(c) During the chase, the core body temperature of both predator and prey rises. Describe how changes in blood circulation help to return their core body temperatures to normal. (3)

☑ **Examiner tip**

In longer questions like this try to be clear on writing cause and effect. Where possible use key terms and concepts from your course as part of your description as you will often receive credit for these. However, the terms need to be in the correct context – you will not gain marks for lists of random terms that do not demonstrate your understanding of what they mean.

Student answer	Examiner comments
An increase in core temperature causes vasodilation so that more heat is lost from the skin.	This response would only score 1 mark for the recognition that more heat would be lost from the skin. The reference to vasodilation is not enough as it does not describe what change occurs to the blood circulation.
This is an example of homeostasis using a negative feedback mechanism. Changes to the core temperature are detected by thermoreceptors in the hypothalamus which send nerve impulses to arterioles in the skin. This causes vasodilation resulting in increased blood flow to the skin.	This response is better because it includes key terms and structures in the correct context of how the change is caused (homeostasis, negative feedback, hypothalamus). It also clearly describes the effect of vasodilation on the blood circulation.

(Edexcel GCE Biology (Salters-Nuffield) Advanced Unit 5 June 2008.)

Practice questions

1 (a) Name the region of the human brain involved in control of heart rate. (1)

(b) Heart rate increases during exercise. Explain the mechanisms involved in controlling this increase in heart rate. (4)

Total 5 marks

(Edexcel GCE Biology (Salters-Nuffield) Advanced Unit 5 June 2007)

2 Doing too little exercise can lead to health problems, but too much exercise can also be harmful. Discuss the benefits and potential dangers of exercise in humans. (6)

Total 6 marks

(Edexcel GCE Biology (Salters-Nuffield) Advanced Unit 5 June 2007)

3 The table below refers to three major stages of aerobic respiration and the products of each stage. Copy and complete the table by inserting the part of the cell in which the stage occurs and two products in the blank spaces.

Stage	Part of cell in which it occurs	Two products
glycolysis		
Krebs cycle	matrix of mitochondrion	
electron transport chain		ATP and water

(4)

Total 4 marks

(Edexcel GCE Biology Advanced Unit 4 – paper 3 June 2007)

4 The diagrams show one sarcomere in its fully relaxed state and when it is partially contracted.

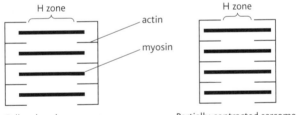

Fully relaxed sarcomere   Partially contracted sarcomere

(a) Calculate the percentage change in width of the H zone when the sarcomere is partially contracted. Show your working. (3)

(b) During the contraction of this sarcomere, the myosin filaments pull the actin filaments towards the centre of the sarcomere. Explain how this is brought about. (4)

Total 7 marks

(Edexcel GCE Biology Advanced Unit 4 – paper 3 June 2007)

5 The diagram shows the ways in which the respiratory system and different parts of the brain interact with each other to regulate breathing.

Cerebral hemispheres

Respiratory centres in the pons and medulla

Intercostal muscles and stretch receptors

Chemoreceptors

Diaphragm muscles and stretch receptors

(a) Breathing can be controlled voluntarily and involuntarily. Name the part of the brain that controls **involuntary** breathing (1)

(b) Suggest *one* occasion when the depth of breathing is increased voluntarily. (1)

(c) Using the information in the diagram, explain the roles of **muscle spindles** and **nerves** in the control of breathing during exercise. (3)

(d) The ventilation of the lungs during breathing is essential in maintaining the concentration gradients of the respiratory gases. This ensures that gas exchange is efficient. Explain why the **chemoreceptors** are particularly important during exercise. (2)

Total 7 marks

(Edexcel GCE Biology Advanced Unit 4 – paper 3 June 2007)

6   The diagram shows some of the muscles in a human leg.

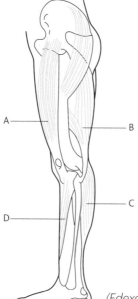

(a) Using the letters **A**, **B**, **C** or **D**, identify the muscle on the diagram above which **(i)** contracts to bend the leg backwards at the knee AND **(ii)** is antagonistic to the muscle identified in (i). (1)

(b) Joint injuries often shorten the career of athletes. Explain the advantages of keyhole surgery on damaged joints, such as the knee, compared with traditional surgery. (2)

(c) Two weeks after taking part in a 56 km race, 33% of the runners developed respiratory tract infections. Those who completed the race were three times more likely to develop an infection after the race compared with a control group which did not run.

Explain *one* factor which could contribute to this higher infection rate. (3)

Total 6 marks

(Edexcel GCE Biology (Salters-Nuffield) Advanced Unit 5 June 2005)

7   The diagram shows the pathways for the conduction of electrical impulses during the cardiac cycle.

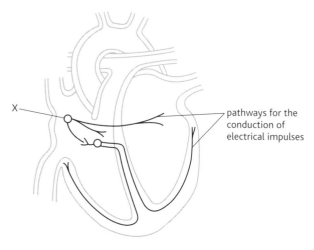

pathways for the conduction of electrical impulses

(a) Name the structure labelled X. (1)

(b) Describe how the structures shown in the diagram control the complete cardiac cycle. (4)

Total 5 marks

(Edexcel GCE Biology (Salters-Nuffield) Advanced Unit 1 January 2005)

Responding to the environment

The nervous system in animals

Animal nervous systems are fast-acting communication systems containing nerve cells (**neurones**) which carry information in the form of nerve impulses (see page 62).

In mammals **sensory neurones** carry impulses from receptors to a central nervous system (CNS) consisting of the brain and spinal cord. The CNS (containing relay neurones) processes information from many sources and then sends out impulses via motor neurones to effector organs (mainly muscles and glands).

The pupil reflex

The iris contains pairs of antagonistic muscles (radial and circular muscles) that control the size of the iris under the influence of the autonomic nervous system (involuntary).

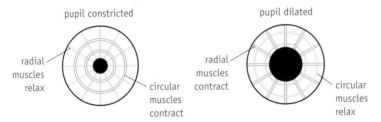

How the muscles of the iris act to control the amount of light entering the eye.

In high light intensities photoreceptors such as **rods** in the retina cause nerve impulses to pass along the **optic nerve** to a group of nerve cells in the brain. These then send impulses along **parasympathetic** motor neurones to the **circular muscles** of the iris. The muscles contract, reducing the diameter of the pupil so that less light can enter the eye, thus preventing damage to the retina.

In low light conditions fewer impulses reach the coordinating centre in the brain – impulses are sent down **sympathetic** motor neurones to the radial muscles of the iris instead. This causes the **radial muscles** to contract and the pupil becomes dilated, allowing more light to reach the retina.

Sensitivity in plants

- **Photoperiodism:** Plants flower and seeds germinate in response to changes in day length. The **photoreceptor** involved is a blue-green pigment called **phytochrome**. On absorbing natural (or red) light phytochrome converts from the inactive form P_R to the active form P_{FR}. In the dark P_{FR} slowly reverts back to P_R because it is relatively unstable (or it can change back rapidly into P_R if exposed to far red light). It is thought that the active P_{FR} may trigger a range of different photoperiodic responses.
- **Phototropism:** Tropisms are growth responses in plants where the direction of the growth response is determined by the direction of the external stimulus. If a plant grows towards a stimulus it is said to be a positive tropic response.

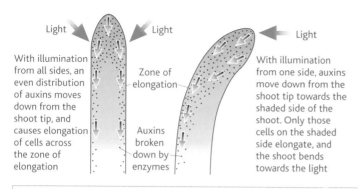

With illumination from all sides, an even distribution of auxins moves down from the shoot tip, and causes elongation of cells across the zone of elongation

Light Light Light

Zone of elongation

Auxins broken down by enzymes

With illumination from one side, auxins move down from the shoot tip towards the shaded side of the shoot. Only those cells on the shaded side elongate, and the shoot bends towards the light

It is not clear what the receptor for phototropism is in shoots, but a good candidate in cereals is riboflavin. The effector for the growth response is cell elongation. This happens just below the tip of the shoot and is controlled by the plant growth substance **IAA** (the first **auxin** discovered).

Mechanism of phototropism in shoots.

Nervous system in mammals	Endocrine system in mammals	Tropisms in plants
Electrochemical changes giving an electrical impulse. Chemical neurotransmitters used at most synapses.	Chemical hormones from endocrine glands carried in the blood plasma around the circulatory system.	Chemical growth substances (e.g. auxins) diffusing from cell to cell – some may go in the plant transport system – the phloem.
rapid acting	slower acting	slower acting
Usually associated with short-term changes, e.g. muscle contraction.	Can control long-term responses, e.g. growth and sexual development. Some are involved in homeostasis, e.g. control of blood sugar. Some can be relatively fast, e.g. effects of adrenaline in response to stress.	Controls long-term growth responses, e.g. cell elongation.
Response is very local and specific such as a muscle cell or gland.	Response may be widespread, or restricted to specific target cells.	Response may be widespread, but normally restricted to cells within a short distance of the growth substance being released.

Table to compare communication and coordination methods in plants and animals.

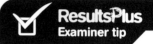

Quick Questions

Q1 Explain what is meant by the term photoreceptor.

Q2 Explain why it is an advantage that shoots have positive phototropism and roots have negative phototropism.

Q3 What effect does IAA have on cells?

Thinking Task

Q1 Why is it an advantage for animals to have a nervous system and an endocrine system?

Neurones and nerve impulses

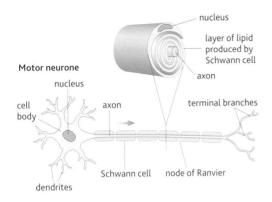

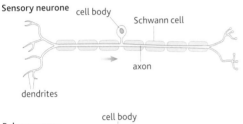

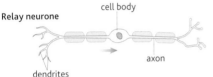

The structure of neurones in a mammal.

The structure of neurones

All **neurones** (nerve cells) have a **cell body** (containing the nucleus and most of the cell's organelles within the cytoplasm), **dendrites** (that conduct impulses towards the cell body) and an **axon** (that conducts impulses away from the cell body). The main difference between the structures of sensory, motor and relay neurones is the relative position of the cell body.

Neurones are able to carry waves of electrical activity called **action potentials** (nerve impulses) over a long distance because the axons can be very long and the membranes are **polarised** (different charges on the inside and outside of the membrane).

Most vertebrate neurones have a fatty insulating layer called the **myelin sheath** wrapped around the axon and/or dendron. This increases the speed of conduction of a nerve impulse through a process called **saltatory conduction**. **Schwann cells** wrap around the neurone, to nourish and protect it and produce the myelin sheath. However, there are small gaps left uncovered called the **nodes of Ranvier**. Action potentials jump from one node of Ranvier to the next, increasing the speed of conduction.

The transmission of a nerve impulse

In a resting neurone there are more **sodium ions** outside the cell membrane than inside, and more **potassium ions** inside than outside. The inside of the resting neurone has a negative charge in comparison to the outside due to the presence of chloride ions and negatively charged proteins. This distribution of ions creates a **potential difference** (a difference in charge) of about $-70\,mV$ called the **resting potential** and the membrane is said to be **polarised**. The **sodium–potassium pump** creates concentration gradients across the membrane (sodium moves out and potassium moves in to the axon). **Potassium ion channels** allow **facilitated diffusion** of potassium ions back out of the membrane down their concentration gradient, creating the uneven distribution of charge across the membrane.

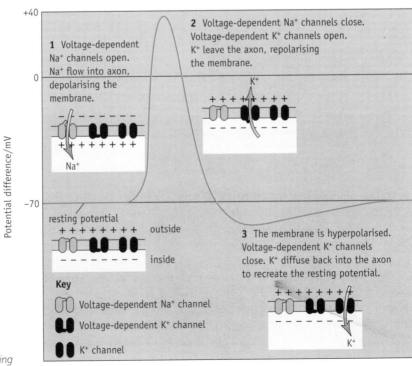

Movement of ions in and out of the membrane during an action potential.

If a neurone cell membrane is stimulated, **voltage-dependent sodium ion channels** open and sodium ions diffuse in. This increases the positive charge inside the cell, so the charge across the membrane is reversed. The membrane now carries a potential difference of about +40 mV. This is the **action potential** and the membrane is said to be **depolarised**. As the charge reverses, the sodium ion channels shut and **voltage-dependent potassium** ion channels open so that more potassium ions leave the axon, repolarising the membrane.

Propagation of a nerve impulse along an axon

At resting potential there is positive charge on the outside of the membrane and negative charge on the inside, with high sodium ion concentration outside and high potassium ion concentration inside.

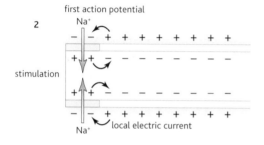

When stimulated, voltage-dependent sodium ion channels open, and sodium ions flow into the axon, depolarising the membrane. Localised electric currents are generated in the membrane. Sodium ions move to the adjacent polarised (resting) region causing a change in the electrical charge (potential difference) across this part of the membrane.

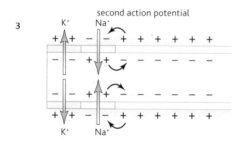

The change in potential difference in the membrane adjacent to the first action potential initiates a second action potential. At the site of the first action potential the voltage-dependent sodium ion channels close and voltage-dependent potassium ion channels open. Potassium ions leave the axon, repolarising the membrane. The membrane becomes hyperpolarised.

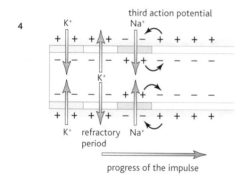

A third action potential is initiated by the second. In this way local electric currents cause the nerve impulse to move along the axon. At the site of the first action potential, potassium ions diffuse back into the axon, restoring the resting potential.

Propagation of a nerve impulse along an axon.

Action potentials have an **all-or-nothing** nature (the values of the resting and action potentials are always the same for a specific neurone). A bigger stimulus increases the frequency of the action potentials (not the strength). A **threshold stimulus** must be applied to produce an action potential.

Straight after an action potential there is a short **refractory period** when a new action potential can't be generated because the sodium ion channels can't reopen. This ensures that action potentials pass along as separate signals and are **unidirectional** (only able to pass in one direction).

Synapses

The point where one neurone meets another is called a **synapse**. At the tip of the axon an impulse opens **calcium ion channels** then triggers the release of a chemical **neurotransmitter** (for example **acetylcholine**) from synaptic vesicles. The neurotransmitter can diffuse across the gap between the neurones (the synaptic cleft) and bind to receptors on the postsynaptic membrane. If the neurotransmitter comes from an excitatory neurone it may open sodium ion channels on the postsynaptic membrane, triggering a new action potential in the postsynaptic neurone. However, some neurotransmitters are inhibitory and they may open chloride ion channels on the postsynaptic membrane, causing it to become hyperpolarised and therefore harder to get an above-threshold response needed to trigger the new action potential.

An enzyme is often present in the synaptic cleft to hydrolyse the neurotransmitter, so the response does not keep happening. The neurotransmitter may also be taken back up into the presynaptic membrane ready to be used again.

Because the receptors are only on one side of the synapse (the postsynaptic membrane) the signal can only pass in one direction (**unidirectional**). Synapses also act as junctions and allow nerve impulses to converge or diverge because one neurone can meet many others at a single synapse.

ResultsPlus
Examiner tip

Pre – before; post – after; uni – one; summation – adding.

If two or more excitatory impulses arrive at a synapse at the same time their effect will be combined and you are more likely to depolarise the postsynaptic membrane (this is spatial summation). If you have a strong stimulus along one neurone many action potentials will arrive one after the other (due to the high frequency) and this will have the same effect (this is temporal summation).

1 An action potential arrives.

2 The membrane depolarises. Calcium ion channels open. Calcium ions enter the neurone.

3 Calcium ions cause synaptic vesicles containing neurotransmitter to fuse with the presynaptic membrane.

4 Neurotransmitter is released into the synaptic cleft.

5 Neurotransmitter binds with receptors on the postsynaptic membrane. Cation channels open. Sodium ions flow through the channels.

6 The membrane depolarises and initiates an action potential.

7 When released the neurotransmitter will be taken up across the presynaptic membrane (whole or after being broken down), or it can diffuse away and be broken down.

axon — synaptic vesicle — neurotransmitter — presynaptic membrane — synaptic cleft — postsynaptic membrane — Ca^{2+} — Na^+

The sequence of events occurring when an action potential arrives at a synapse.

Thinking Task

Q1 Look back at your AS Biology notes for the structure of membranes and transport across membranes. Sketch a diagram of a membrane to show how sodium and potassium ions move during a nerve impulse. Then add to your diagram any other ways that substances can move across membranes.

Quick Questions

Q1 Explain the difference between depolarisation and hyperpolarisation.

Q2 How do the structure of the synapse and axon membrane ensure that nerve impulses are only able to travel in one direction?

Q3 Describe what happens to sodium ions when a neurone membrane is stimulated.

Vision

Receptors are specialised cells able to detect **stimuli**. Receptors are often grouped together into **sense organs**.

Human photoreceptors

Human eyes have two types of **photoreceptor** cells found in the **retina** on the back of the eye. Cones allow colour vision in bright light and are clustered in the centre of the retina. **Rods** only provide black and white vision, but are much more sensitive than cones and can work in dim light conditions.

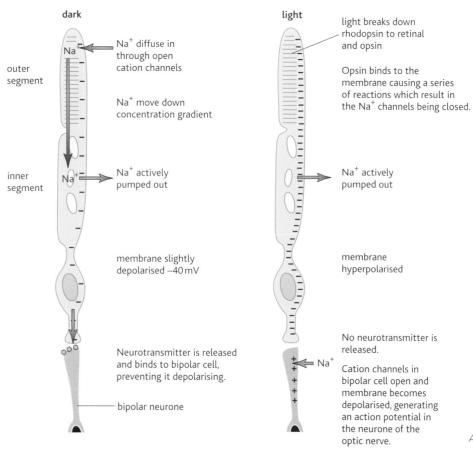

dark

light

light breaks down rhodopsin to retinal and opsin

outer segment

Na$^+$ diffuse in through open cation channels

Opsin binds to the membrane causing a series of reactions which result in the Na$^+$ channels being closed.

Na$^+$ move down concentration gradient

inner segment

Na$^+$ actively pumped out

Na$^+$ actively pumped out

membrane slightly depolarised −40 mV

membrane hyperpolarised

Neurotransmitter is released and binds to bipolar cell, preventing it depolarising.

No neurotransmitter is released.

Cation channels in bipolar cell open and membrane becomes depolarised, generating an action potential in the neurone of the optic nerve.

bipolar neurone

A rod cell in the dark and in the light.

Light energy is absorbed by rhodopsin which splits into retinal and opsin. The opsin binds to the membrane of the outer segment of the cell and this causes sodium ion channels to close. The inner segment continues to pump sodium ions out of the cell and the membrane becomes hyperpolarised (more negative). This means that glutamate is not released across the synapse. Glutamate usually inhibits the neurones which connect the rod cells to the neurones in the optic nerve. When there is less inhibition an action potential forms and is transmitted to the brain. The information from the optic nerve is processed by the brain in the visual cortex.

Quick Questions

Q1 Explain why rods release a neurotransmitter in the dark, but not in the light.

Q2 Describe what happens to rhodopsin when it is exposed to light.

Q3 Compare photoreceptors in mammals and plants.

Thinking Task

Q1 Groups of three rod cells connect to a single bipolar cell whereas just one cone cell connects to a bipolar cell. Use this information to explain why you can't see colour well in dim light conditions.

The structure of the human brain

The cerebrum

The **cerebrum** (cerebral cortex) is the largest part of the brain. It is divided into two **cerebral hemispheres** connected by a band of white matter called the corpus callosum. The cerebrum is associated with advanced mental activity like language, memory, calculation, processing information from the eyes and ears, emotion and controlling all of the voluntary activities of the body.

Frontal lobe (also referred to as the higher centres of the brain) – concerned with the higher brain functions such as decision making, reasoning, planning and consciousness of emotions. It is also concerned with forming associations (by combining information from the rest of the cortex) and with ideas. It includes the primary motor cortex which has neurones that connect directly to the spinal cord and brain stem and from there to the muscles. It sends information to the body via the motor neurones to carry out movements. The motor cortex also stores information about how to carry out different movements.

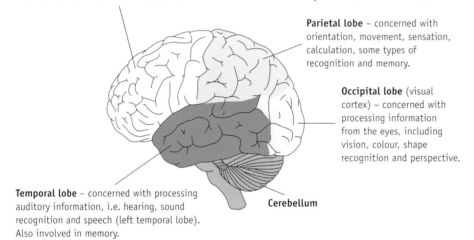

Parietal lobe – concerned with orientation, movement, sensation, calculation, some types of recognition and memory.

Occipital lobe (visual cortex) – concerned with processing information from the eyes, including vision, colour, shape recognition and perspective.

Cerebellum

Temporal lobe – concerned with processing auditory information, i.e. hearing, sound recognition and speech (left temporal lobe). Also involved in memory.

The regions of the cerebral hemispheres and their functions.

The cerebellum and brain stem

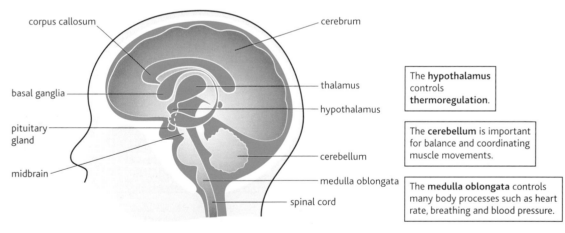

corpus callosum

cerebrum

basal ganglia

thalamus

pituitary gland

hypothalamus

midbrain

cerebellum

medulla oblongata

spinal cord

The **hypothalamus** controls **thermoregulation**.

The **cerebellum** is important for balance and coordinating muscle movements.

The **medulla oblongata** controls many body processes such as heart rate, breathing and blood pressure.

The main regions of the human brain.

⟨?⟩ Quick Questions

Q1 Distinguish between the cerebrum and the cerebellum.

Q2 Which region of the brain is most associated with thinking and decision making?

Brain development

We are born with a range of **innate behaviours** (behavioural responses that do not need to be learnt) such as crying, grasping and sucking. However, the brain still needs much growth and development after birth through the formation of synapses and the growth of axons.

Evidence for critical windows

Critical windows (or critical periods) for development are those periods of time where it is thought that the nervous system needs specific stimuli in order to develop properly.

Evidence for critical windows for development has come from **medical observations** (e.g. children who develop cataracts before the age of 10 days may suffer from permanent visual impairment even if the cataracts are repaired at a later date) and from **animal models**. Hubel and Wiesel used kittens and monkeys as models to investigate the critical window in visual development because of the similarity of their visual systems to that of humans.

The animals were deprived of the stimulus of light into one eye (monocular deprivation) at different stages of development and for different lengths of time. They found that kittens deprived of light in one eye at 4 weeks after birth were effectively permanently blind in that eye. Monocular deprivation before 3 weeks and after 3 months had no effect. It was thought that during the critical period (about 4 weeks after birth) connections to cells in the visual cortex from the light-deprived eye had been lost. This meant that the eye that remained open during development became the only route for visual stimuli to reach the visual cortex.

Eye deprived of light during critical window	Eye that remains open during the critical window
Axons do not pass nerve impulses to cells in the visual cortex.	Axons pass nerve impulses to cells in the visual cortex.
Inactive synapses are eliminated.	Synapses used by active axons are strengthened.
Eye has no working connection to the visual cortex and is effectively blind, even though the cells of the retina and optic nerve work normally when exposed to light.	Synapses only present for axons coming from the light-stimulated eye. So the visual cortex can only respond to this eye.

Issues about the use of animals for research

The use of animals as models for understanding how humans develop, or how new drugs may affect us, is a very controversial area. There are those who hold an absolutist view of **animal rights** and think we should *never* keep animals or use them in medical research. From the point of view of medical research, a much more widespread position is the relativist view that humans should treat animals well and minimise harm and suffering so far as is possible. Here the emphasis is on **animal welfare**, respecting their rights to such things as food, water, veterinary treatment and the ability to express normal behaviours. This is pretty much the position in European law. This all assumes that animals can suffer and experience pleasure.

A utilitarian ethical framework allows certain animals to be used in medical experiments *provided* the overall expected benefits are greater than the overall expected harms based on the belief that the right course of action is the one that maximises the amount of overall happiness or pleasure in the world.

ResultsPlus
Examiner tip

Visual development is an example of how the effects of nature and nurture can combine in development. The genes control the development of the responsive cells in the visual cortex (nature) but a stimulus from the environment is needed during the critical window for the correct connections to be made (nurture).

The role of nature and nurture in brain development

- **Nature:** Many of our characteristics develop solely under the influence of our genes with little influence from our environment or learning, e.g. blood group.

- **Nurture:** Many characteristics are learnt or are heavily influenced by the environment, e.g. how long your hair is.

Most of our characteristics are actually determined by nature and nurture or nature via nurture. We are the result of a mixture of genetic and environmental factors. Human behaviours, attitudes and skills may have an underlying genetic basis but are modified by experience or the environment in a way which is very complex. For example, the chance of developing some diseases, such as some cancers, has a genetic basis, where a gene or several genes interact to confer susceptibility to the disease with environmental factors contributing to the risk of developing the disease.

Evidence for the relative roles of nature and nurture in brain development come from a variety of sources:

- **The abilities of newborn babies:** Newborn babies have some innate capacities. These suggest that genes help to form the brain and some behaviours before the baby is born.

- **Studies of individuals with damaged brain areas:** Some patients who have suffered from brain damage show the ability to recover some of their brain function. This demonstrates that some neurones have the ability to change.

- **Animal experiments:** e.g. Hubel and Weisel's experiments on critical windows for sight, suggest that external stimulation is important in brain development.

- **Twin studies:** Identical twins share all the same genes. Fraternal (non-identical) twins share the same number as any other sibling would. Twin studies can help to estimate the relative contribution of genes and the environment. Any differences between identical twins must be due to the effects of the environment.

  Identical twins raised apart in comparison to those raised together are particularly useful for study. For example if there is a greater difference between those twins raised apart than twins raised together it suggests some environmental influence. However, twins raised apart may not have completely different environments and twins raised together may develop different personalities due to a desire to be different. In general if genes have a strong influence on the development of a characteristic, then the closer the genetic relationship, the stronger the correlation will be between individuals for that trait.

- **Cross-cultural studies:** Investigations into the visual perception of groups from different cultural backgrounds support the idea that visual cues for depth perception are at least partially learnt.

⌖ Thinking Task

Q1 What is your personal view on the use of animals in medical research? For example, how many **a** fruit flies, **b** mice, **c** cats, **d** monkeys do you think you could use to test new drugs to help treat **i** breast cancer, **ii** malaria, **iii** wrinkles in the skin?

How do you justify your position?

⟨?⟩ Quick Questions

Q1   Describe why it may be dangerous to leave a patch over the damaged eye of a child for a prolonged period of time.

Q2   Explain why kittens and monkeys have been used in experiments looking at human brain development.

Q3   If one identical twin has schizophrenia there is 80% chance that their twin will also have symptoms of schizophrenia. However, if one fraternal twin has schizophrenia there is only a 15% chance that their twin will also have symptoms of schizophrenia. What do these figures suggest about the contribution of nature and nurture on the development of schizophrenia?

Learning and habituation

Learning is a process that results in a change in behaviour (or knowledge) as a result of experience. For learning to be effective you must remember what you have learnt. Memories (conscious and sub-conscious) are formed by changing or making new synapses in the nervous system.

Habituation

Habituation is a very simple type of learning that involves the loss of a response to a repeated stimulus which fails to provide any form of **reinforcement** (reward or punishment). It allows animals to ignore unimportant stimuli so that they can concentrate on more rewarding or threatening stimuli.

Investigating habituation

The core practical is an example of a simple investigation into habituation. It measures the time a snail spends withdrawn into its shell when you tap the surface it is moving on at regular time intervals or gently touch the snail's head. Initially the snail tends to retreat into its shell for a significant period of time after each tap. As the tapping continues the snail stays in its shell for a shorter duration as it becomes habituated to the tapping.

Many invertebrates have been useful **animal models** for investigating the workings of the nervous system. For example sea slugs (*Aplysia*) have been used to investigate habituation.

A Gill withdraws when siphon stimulated.

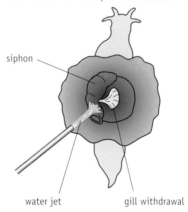

siphon

water jet   gill withdrawal

B After several minutes of repeated stimulation of the siphon the gill no longer withdraws.

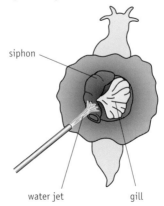

siphon

water jet   gill

Habituation in a sea slug.

C How habituation is achieved.

1 With repeated stimulation, Ca^{2+} channels become less responsive so less Ca^{2+} crosses the presynaptic membrane.

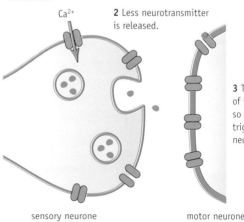

Ca^{2+}

2 Less neurotransmitter is released.

3 There is less depolarisation of the postsynaptic membrane so no action potential is triggered in the motor neurone.

sensory neurone from the siphon

motor neurone to the gill

ResultsPlus
Examiner tip

Describing how to investigate habituation to a stimulus is a required practical so you may well be asked questions about this during the exam. As this is an experiment involving animals (possibly humans, depending on your method) you should consider any ethical and safety issues that may arise in your methodology. It is also worth considering how to evaluate your results as it is often difficult to control many variables when using live animals in experiments.

Quick Questions

Q1 Write out the reflex arc involved in the sea slug's response to water being sprayed onto its siphon.

Q2 Suggest why sea slugs used in this habituation experiment need to have been reared in captivity rather than in the sea.

Q3 Suggest whether nature or nurture is likely to be responsible for the development of an innate reflex.

Effects of imbalances in brain chemicals

Dopamine and Parkinson's disease

Parkinson's disease is associated with the death of a group of dopamine-secreting neurones in the brain (an area of the midbrain known as the substantia nigra). This results in the reduction of dopamine levels in the brain. Dopamine is a **neurotransmitter** which is active in neurones in the **frontal cortex**, brain stem and spinal cord. It is associated with the control of movement and emotional responses.

The symptoms of Parkinson's are:
- muscle tremors (shakes)
- stiffness of muscles and slowness of movement
- poor balance and walking problems
- difficulties with speech and breathing
- depression.

A variety of treatments are available for Parkinson's disease, most of which aim to increase the concentration of dopamine in the brain. Dopamine cannot move into the brain from the bloodstream, but the molecule which is used to make dopamine can. This molecule is called **L-dopa** (levodopa) and can be turned into dopamine to help control the symptoms. Some other treatments for Parkinson's are outlined later in this section.

Serotonin and depression

Serotonin is a neurotransmitter linked to feelings of reward and pleasure. A lack of serotonin is linked to **clinical depression** (prolonged feelings of sadness, anxiety, hopelessness, loss of interest, restlessness, insomnia, etc.).

Treatments for depression often involve drugs that can help increase the concentration of serotonin in the synapses. For example, Prozac is a selective serotonin reuptake inhibitor (SSRI) that blocks the process which removes serotonin from the synapse. See below for discussion on how SSRIs might work.

The effect of drugs on synapses

Many drugs affect the nervous system by interfering with the normal functioning of a **synapse**. The diagram and following text show some of the ways synapses can be affected by drugs.

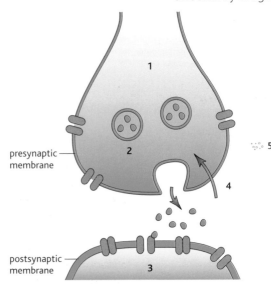

Different stages in synaptic transmission that can be affected by drugs.

1 Some drugs affect the synthesis or storage of the **neurotransmitter**. For example **L-dopa** used in the treatment of Parkinson's disease is converted into **dopamine**, increasing the concentration of dopamine to reduce the symptoms of the disease.

2 Some drugs may affect the release of the neurotransmitter from the **presynaptic** membrane.

3 Some drugs may affect the interaction between the neurotransmitter and the **receptors** on the **postsynaptic** membrane.

   a) Some may be stimulatory by binding to the receptors and opening the **sodium ion channels** – for example dopamine agonists (which mimic dopamine because they have a similar shape and are used in the treatment of Parkinson's disease) bind to dopamine receptors and trigger **action potentials**.

   b) Some may be inhibitory, blocking the receptors on the postsynaptic membranes and preventing the neurotransmitters binding.

4   Some drugs prevent the reuptake of the neurotransmitter back into the presynaptic membrane. For example **ecstasy** (**MDMA**) works by preventing the reuptake of **serotonin**. The effect is the maintenance of a high concentration of serotonin in the synapse which brings about the mood changes in the users of the drug. One of the many possible side effects of ecstasy use is depression as a result of the loss of serotonin from the neurones because of the lack of reuptake. Prozac is a common example of a selective serotonin reuptake inhibitor (SSRI) that blocks the reuptake of serotonin in the treatment of depression.

5   Some drugs may inhibit the enzymes involved in breaking down the neurotransmitter in the synaptic cleft, resulting in the maintenance of a high concentration of the neurotransmitter in the synapse and therefore repeated action potentials (or inhibition) of the postsynaptic neurone.

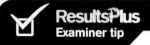

ResultsPlus
Examiner tip

Look back at your notes on synapses and nerve impulses to help get your head around this section about how drugs can affect your nervous system.

The development of new drugs

As we have seen in this section, chemicals that affect membrane-bound proteins or mimic the effect of naturally occurring neurotransmitters can have a significant effect on defective or normal neural pathways. The more we know about the specific proteins (and their shapes) active in cells the more likely we are to find complementary chemicals that can have the desired effect.

Traditionally most medicines are developed from existing chemicals (often extracted from plants), but the information coming from the human genome project (see next page for further details) could help develop drugs that are highly specific so that they can be effective in lower doses with fewer side effects. Pharmacogenomics links pharmaceutical expertise (drug development and manufacture) with the knowledge of the human genome. New drugs have to go through a rigorous process of testing, including animal trials and clinical trials, long before they can reach the market.

Imaging techniques for the brain

Several imaging techniques are useful for medical diagnosis and investigating brain structure and function. For example, the effects of drugs and diseases such as Parkinson's on the activity of the brain can now be seen using imaging techniques such as fMRI.

Magnetic resonance imaging (**MRI**) scans use a magnetic field and radio waves to make images of soft tissues like the brain. MRI scans can be used in the diagnosis of tumours, strokes, brain injuries and infections. They can also be used to track degenerative diseases like Alzheimer's by comparing scans over a period of time.

Functional magnetic resonance imaging (**fMRI**) is a modified MRI technique that can allow you to see the brain in action during live tasks, because it detects activity in the brain by following the uptake of oxygen in active brain areas.

Computerised axial tomography (**CT** or **CAT**) scans use thousands of narrow beam X-rays rotated around the patient. Like MRI they only capture one moment in time and so only look at structures and damage rather than functions. The resolution is worse than MRI so small structures in the brain can't be distinguished; they also use potentially harmful X-rays.

Quick Questions

Q1   Explain why people suffering from Parkinson's may suffer from depression.

Q2   Explain how L-dopa may reduce the symptoms of Parkinson's disease.

Q3   Suggest how a new drug developed to be a similar shape to serotonin may help treat clinical depression.

Thinking Task

Q1 Explain why treating mental health problems with drugs is such a difficult process to get right. (Remember that neurotransmitters are effective in extremely low concentrations and are active in very specific synapses within the brain.)

Uses of genetic modification

The Human Genome Project and the development of new drugs

A genome is all of the DNA (or genes) of an organism. The Human Genome Project was a multinational project that determined the base sequence of the human genome. Many new genes have been identified, including some of those genes responsible for inherited diseases. In addition new **drug targets** (specific molecules that drugs interact with to have their effects, e.g. enzymes) have been identified. Information about a patient's genome may help doctors to prescribe the correct drug at the correct dose. The Human Genome Project may also allow some diseases to be prevented. If you understand what genes you carry you may understand what disease you are likely to be at risk from.

The Human Genome Project also helps to provide information about evolution and increases our knowledge of physiology and cell biology.

Some ethical questions about possible developments from the Human Genome Project

- Who owns the information? Some groups have applied for patents on genetic sequences so that they have ownership, or have to be paid for any treatments developed using the knowledge of that sequence.

- Who is entitled to know the information about your genome if it is sequenced? Should insurance companies have access to the information?

- Will genetic screening lead to eugenics (the genetic selection of humans) and designer babies?

- Who will pay for the development of the new therapies and drugs? Many possible highly specialised treatments may be very expensive and will only be suitable for a few people.

The use of genetically modified (GM) organisms to produce drugs

GM plants may be useful for producing edible drugs such as vaccines that can be stored and transported easily in plant products such as bananas or potatoes. Useful genes can be transferred into crop plants using a **vector** such as *Agrobacterium tumefaciens*, gene guns (pellets coated with DNA) or a virus. **Restriction enzymes** are used to cut DNA at specific sequences and **DNA ligase** is an enzyme that can be used to stick pieces of DNA together. These make it possible to insert specific DNA sequences into the GM organism. Large numbers of identical GM plants can easily be produced.

Transgenic animals (animals with a human gene added to them) can be used to produce useful drugs that can be harvested from their milk (or even semen). Liposomes and viruses are **vectors** used to insert genes into animal cells. Drugs produced from transgenic animals include the blood clotting factors used to treat haemophilia.

Micro-organisms such as bacteria are the most common target for genetic modification as they are relatively easy targets for gene transfer and can be grown rapidly in large quantities in fermenters. The drugs produced can be extracted and purified using downstream processing. Insulin, to treat type II diabetes, is an example of a drug produced from genetically modified micro-organisms.

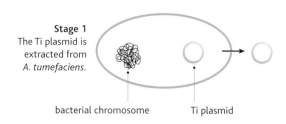

Stage 1
The Ti plasmid is extracted from *A. tumefaciens*.

bacterial chromosome   Ti plasmid

Ti plasmid

Stage 2
The gene to be carried to the plant is inserted into the Ti plasmid which is then returned to the bacterium.

new gene

Stage 3
The plant is infected with the modified bacterium and part of the Ti plasmid with the engineered gene becomes part of the plant chromosomes.

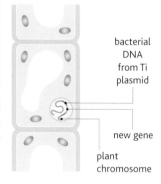

bacterial DNA from Ti plasmid

new gene

plant chromosome

Stage 4
A. tumefaciens causes a tumour to develop on the plant. These plant cells contain the new gene. If tumour cells are taken and cultured, whole new plants can be grown from them, containing the new genes. These are genetically engineered or transgenic plants.

crown gall (tumour) caused by *A. tumefaciens*

new plant containing new gene grown from gall cells

The genetic modification of plants.

Some possible concerns over the development and use of genetically modified organisms (GMOs)

- genetic pollution (transfer of the genes to natural, wild species) through cross-pollination

- antibiotic resistance genes are used to identify GM bacteria which could lead to antibiotic resistance developing in other microbes

- GM crops could become super-weeds that out-compete other plants and may be resistant to herbicides. They could damage natural food chains, resulting in damage to the environment because they would encourage farmers to use more selective herbicides to kill everything but the crop.

- GM crops may not produce fertile seeds. This prevents farmers collecting seed and replanting, so they need to return to the biotechnology company to buy new seeds for each planting. This could make them too expensive for some farmers.

ResultsPlus
Examiner tip

When preparing for your A2 Biology exams try to think why things are the way they are and look for links between different areas of the course. This can often help you understand, remember and apply your knowledge even in areas of the course which may appear tough.

Quick Questions

Q1 Describe what is meant by the term 'genetic pollution of the environment'.

Q2 Describe the benefits of using bacteria to produce a human protein (like insulin) to treat a disease.

Thinking Task

Q1 Outline some of the benefits and disadvantages of setting up a national screening programme for a newly identified gene responsible for an inherited genetic disease.

Topic 8 – Grey matter checklist

By the end of this topic you should be able to:

Revision section	Checkpoints	Spec. point	Revised		Practice exam questions	
Responding to the environment	Explain how the nervous system allows us to respond to the world around us, using the pupil reflex as an example.	LO7		☐		☐
	Describe how plants detect light and respond.	LO2		☐		☐
	Compare plant hormones, animal hormones and the nervous system all as methods of coordination.	LO8		☐		☐
The nervous system and nerve impulses	Describe the structure and function of sensory, relay and motor neurones including the role of Schwann cells and myelination.	LO3		☐		☐
	Describe how a nerve impulse passes along an axon.	LO4		☐		☐
	Describe what synapses do and how they work, including the role of acetylcholine.	LO5		☐		☐
Vision	Describe how the rod cells in the retina work to create action potentials in the optic nerves.	LO6		☐		☐
Structure of the human brain	Recall where the different regions of the human brain are and what each one does. This should include: the cerebral hemispheres, hypothalamus, cerebellum and medulla oblongata.	LO9		☐		☐
Brain development	Discuss the concept of a 'critical window' in the development of vision.	LO11		☐		☐
	Describe Hubel and Wiesel's work with monkeys and kittens and how it explored the development of the brain.	LO12		☐		☐
	Consider the different methods used to study the development of the brain.	LO13		☐		☐
	Discuss two ethical standpoints on the moral and ethical issues relating to the use of animals in medical research.	LO16		☐		☐
Learning and habituation	Describe how animals, including humans, can learn by habituation.	LO14		☐		☐
	Describe how to investigate habituation.	LO15		☐		☐
Effects of drugs on neurotransmitter systems	Explain how chemical levels in the brain may change, resulting in illnesses such as Parkinson's and depression and how this area is a source of research for new drugs.	LO17		☐		☐
	Explain the ways that drugs affect synapses in the brain, including ecstasy and those used to treat Parkinson's.	LO18		☐		☐
	Describe how different imaging techniques are used to study the brain, including magnetic resonance imaging (MRI), functional magnetic resonance imaging (fMRI) and computed tomography (CT) scans.	LO10		☐		☐
Uses of genetic modification	Discuss how the Human Genome Project is helping to develop new drugs and some of the issues that arise.	LO19		☐		☐
	Describe how drugs can be produced using genetically modified organisms (plants and animals and microorganisms).	LO20		☐		☐
	Discuss the risks and benefits of genetically modified organisms.	LO21		☐		☐

ResultsPlus
Build Better Answers

1 People with Parkinson's disease have poor control over their skeletal muscles, caused by a lack of the neurotransmitter dopamine. Large numbers of neurones secreting dopamine are found in the basal ganglia region of the brain.

Parkinson's disease can be diagnosed and monitored using brain scans. The fMRI scans on the right show the results of a study where subjects did a standard finger-tapping activity to investigate the effectiveness of a new drug treatment.

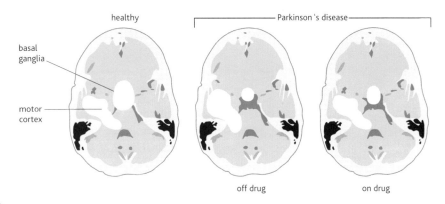

The results above right are from a healthy brain, a patient with Parkinson's disease without drug treatment and a patient with Parkinson's disease taking drug treatment. The scan shows a horizontal section with the front of the head at the top. The most active areas are white.

(a) Using the fMRI scans above, discuss the effects of this new drug on brain activity. (3)

✔ Examiner tip

When provided with plenty of information to read and diagrams to look at, make sure you study it thoroughly to help you understand the context of the question and what the examiner is actually asking you about. For example, this question is comparing activity in different regions of the brain and not the size of the different areas.

Student answer	Examiner comments
There is more activity in the basal ganglia and less activity in the motor cortex for the person treated with the drug than the person with Parkinson's without the drug. The drug may work by stimulating the release of more dopamine from the basal ganglia.	Make sure you make a comparative statement. The student here includes 'more' and 'less' to make the comparison clear. It is also made clear what is being measured by the fMRI – the activity of the brain. Many candidates lost marks for this question by referring to an increase or decrease in the area rather than the activity of the area. This response gains full marks by going on to provide a possible explanation for the differences.

(b) Explain how neurotransmitters, such as dopamine, stimulate neurones. (4)

✔ Examiner tip

Don't get thrown by the context of the example. You may not know much about exactly how dopamine works, but you should be able to recall what happens at synapses and how a neurotransmitter can stimulate an action potential in the next neurone.

Student answer	Examiner comments
Dopamine can bind to receptors on the postsynaptic membrane.	This response describes what dopamine does, but doesn't explain how it stimulates neurones.
Dopamine can be released from vesicles in the presynaptic membrane in response to calcium ions moving in through the membrane when an action potential arrives. The dopamine can diffuse across the synapse and bind to receptors and open sodium ion channels. Sodium ions can enter the postsynaptic membrane and cause the membrane to depolarise, resulting in an action potential in the postsynaptic neurone.	This response provides lots of specific detail about how a neurotransmitter stimulates a new action potential in response to the arrival of an action potential at the synapse.

(Edexcel GCE Biology (Salters-Nuffield) Advanced Unit 5 June 2008.)

Practice questions

1 (a) (i) Describe how the nervous system controls the pupil reflex in a mammal in response to bright light. (4)

(ii) Describe and explain how myelination of neurones is an advantage in this reflex pathway. (3)

(b) Hubel and Weisel covered one eye of kittens of different ages to investigate the timing of visual development in mammals.

Kittens which had one eye covered from the fourth to the fifth week subsequently had very poor vision in that eye. Kittens which had one eye covered at earlier or later times had normal vision. Suggest an explanation for these observations. (3)

(c) Some people have ethical objections to animal experiments. Suggest how a biologist might justify the use of animals in experiments. (2)

Total 12 marks
(Edexcel GCE Biology (Salters-Nuffield) Advanced Unit 5B June 2007)

2 Detection of light occurs in both mammals and flowering plants.
(a) In humans, the central region of the retina has very few rod cells. However, in a dog about 80–90% of the photoreceptors in the central region of the retina are rod cells.

Suggest *one* advantage to a dog of having more rod cells in this region of the retina. (3)

(b) Describe the detection of light in flowering plants. (3)

Total 6 marks
(Edexcel GCE Biology Advanced Unit 4 June 2008 Q3)

3 The diagram below shows a vertical section through a human brain.
Using the letters A, B, C, D or E, state which region of the brain:

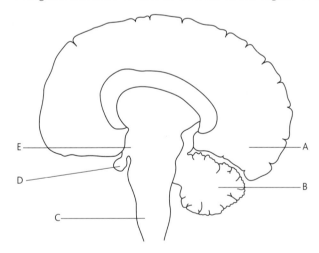

(a) coordinates movement (1)

(b) controls heart rate (1)

(c) receives sensory input from the eyes. (1)

Total 3 marks
(Edexcel GCE Biology (Salters-Nuffield) Advanced Unit 5B June 2005)

4 Twin studies can be used to investigate the role of genes in behaviour. The frequency of schizophrenia in identical (monozygotic) twins was investigated. Conditions such as schizophrenia are thought to be polygenic.

   (a) Define the term polygenic.   (1)

   (b) If one monozygotic twin has schizophrenia then the probability of the second twin having the condition is 46%.

       Predict what you would expect the percentage probability to be if schizophrenia was entirely caused by genes.   (1)

   (c) The probability of two unrelated people both having schizophrenia is 0.5%.

       Explain what the results of this study show about the roles of genes and the environment in schizophrenia.   (2)

Total 4 marks

(Edexcel GCE Biology (Salters-Nuffield) Advanced Unit 5B June 2006)

5 The Human Genome Project has discovered the location of 30 000 genes. Only a small number of human genes have a known function, so the next step is to find out what the rest of the genes do.

   (a) Explain what is meant by the word 'genome'.   (1)

   (b) Some scientists want to use knowledge gained from the Human Genome Project to screen people to find out if they have a genetic predisposition to certain diseases, such as heart disease or lung cancer. They think that screening can help people to lead a healthier life.

       Other scientists think that genetic screening should not be carried out because it will create extra problems for society.

       (i) Suggest how knowing that you were more likely than other people to develop heart disease or lung cancer could help you to lead a longer, healthier life.   (2)

       (ii) Suggest how compulsory genetic screening of everyone might be of benefit to society.   (2)

       (iii) Suggest why people might vote against compulsory genetic screening in a referendum.   (3)

Total 8 marks

(Edexcel GCE Biology (Salters-Nuffield) Advanced Unit 1 June 2004)

Unit 5 specimen paper

1 Muscle paralysis is common in many cases of poisoning, often as a result of interference with chemical transmission from the motor neurones to the muscles at the neuromuscular junctions. Studies of venomous snakes, such as the Prugasti krait (*Bungarus fasciatus*) have played a part in the investigation of this chemical transmission.

   (a) Describe the normal sequence of events that occurs **within a muscle fibre** after stimulation of a neuromuscular junction. (5)

   (b) Bungaratoxin can be isolated from the venom of the Prugasti krait. In minute amounts, it can cause paralysis of the diaphragm and intercostal muscles by its effects at synapses. Suggest how bungaratoxin causes these effects. (3)

   Total 8 marks

   (Edexcel GCE Biology (Salters-Nuffield) Advanced Unit 5B June 2007)

2 Isolated mitochondria in a solution containing inorganic phosphate and an electron donor can be used to study respiration. An electrode is used to record changes in oxygen concentration while mitochondria respire. The graph shows changes in oxygen concentration for some isolated mitochondria.

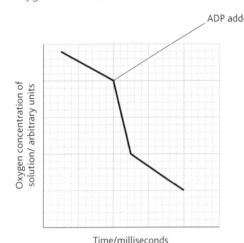

   (a) (i) Describe and explain the trends shown on the graph above. (3)

   (ii) Name an electron donor used in the electron transport chain in mitochondria. (1)

   (iii) State the location of the electron transport chain in mitochondria. (1)

   (iv) Describe how ATP is synthesised in the electron transport chain. (4)

   (b) ATP is used to provide an immediate supply of energy for biological processes. Describe the role of ATP in the following processes.

   (i) nerve impulse transmission (2)

   (ii) hyperpolarisation of rod cells in the retina. (2)

   Total 13 marks

   (Edexcel GCE Biology (Salters-Nuffield) Advanced Unit 5B June 2008)

3 **(a)** Explain what is meant by the Human Genome Project. (2)

   (b) The Human Genome Project is making it possible to identify people who may be at risk of developing medical conditions such as heart disease, cancer and diabetes.

   (i) Suggest *two* reasons why identifying people at risk might be of benefit to the people who are tested. (2)

   (ii) Suggest *three* disadvantages or ethical objections posed by the Human Genome Project. (3)

   Total 7 marks

   (Edexcel GCE Biology (Salters-Nuffield) Advanced Unit 2 June 2005)

4 (a) At high environmental temperatures, the rate of sweating in humans increases. Explain how sweating is involved in the regulation of body temperature. (2)

(b) In an investigation, a healthy volunteer measured his body temperature. After 5 minutes, he got into a bath of water at a temperature of 18 °C. He stayed in the bath for 10 minutes, then got out and sat on a chair. During the investigation, he recorded his body temperature at regular time intervals. The results of this investigation are shown in the table below.

Time / min	Activity	Body temperature / °C
0	Started investigation	37.0
5	Got into bath	36.9
10	Lying in bath	36.7
15	Got out of bath	36.5
20	Sitting on a chair	36.8
25	Sitting on a chair	37.0

(i) Describe the changes in body temperature that occurred during this investigation. (3)

(ii) Suggest explanations for the changes in body temperature that occurred between the following time intervals:
     5 to 10 minutes
     15 to 25 minutes (3)

Total 8 marks

(Edexcel GCE Biology Advanced – paper 6112/01 June 2008)

5 An investigation was carried out into the effect of cycling speed on the breathing rate of a healthy student. In this investigation, an exercise bicycle was used.

The breathing rate of the student was measured at rest. He then cycled at 10 km per hour for 2 minutes and, immediately after, his breathing rate was recorded. He rested for 5 minutes, before cycling at 15 km per hour for 2 minutes, after which his breathing rate was again measured.

This investigation was repeated at cycling speeds of 20 and 25 km per hour. The student rested for 5 minutes between each period of cycling. The results are shown in the table below.

Cycling speed / $km\,h^{-1}$	Breathing rate / breaths min^{-1}
0 (rest)	12
10	14
15	17
20	20
25	27

(a) Calculate the percentage increase in breathing rate, as the cycling speed increased from 10 km per hour to 25 km per hour. Show your working. (2)

(b) Suggest an explanation for these results. (2)

Total 4 marks

(Edexcel GCE Biology Advanced – paper 6112/01 June 2008)

6 (a) The diagram shows some of the stages of anaerobic respiration in a muscle cell.

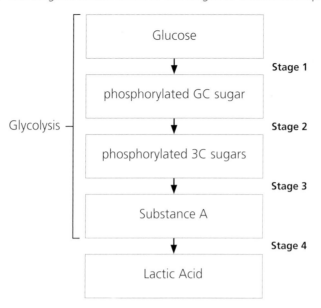

(i) Name substance A. (1)

(ii) State which of the stages shown in the diagram:

1. uses ATP

2. produces ATP. (2)

(b) The Krebs cycle occurs during aerobic respiration and is an example of a metabolic pathway.

(i) Explain why the Krebs cycle is described as a metabolic pathway. (1)

(ii) State precisely where in the cell the Krebs cycle occurs. (1)

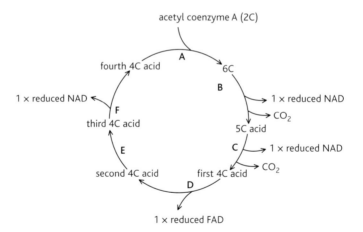

(c) The diagram shows some of the stages that occur in the Krebs cycle. Oxidoreductase enzymes are involved in some of the reactions in the Krebs cycle.

Using the letters A to F and the information given in the diagram, list *all* the stages that involve an oxidoreductase enzyme. (1)

Total 6 marks

(Edexcel GCE Biology Advanced Unit 4 – paper 3 June 2008)

7 (a) The diagram below shows a spirometer. This apparatus is used to measure the volume of air breathed in and out and the frequency of breathing under different conditions.

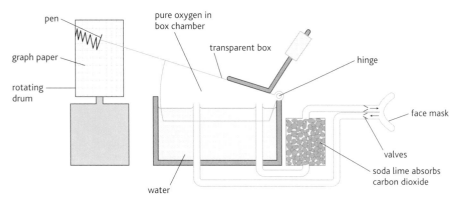

A spirometer was used to compare a person's breathing at rest and during exercise. The results are shown in the graphs below.

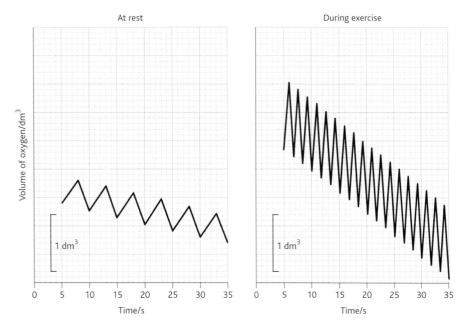

The minute volume is the volume of oxygen taken into the lungs in 1 minute, and is calculated by multiplying the tidal volume by the breathing rate.

(i) Using the information on the graphs, calculate the minute volume at rest. Show your working. (2)

(ii) Calculate the increase in the minute volume that occurred in this person as a result of the exercise. Show your working. (2)

(b) Cardiac output also increases during exercise.

(i) State what is meant by cardiac output. (1)

(ii) Explain how increases in minute volume and cardiac output during exercise enable rapid delivery of oxygen to muscles. (2)

Total 7 marks

(Edexcel GCE Biology (Salters-Nuffield) Advanced Unit 5B June 2004)

Unit 5 – Comprehension practice

ResultsPlus
Examiner tip

When preparing for the comprehension section of the Unit 5 exam:

- Read the article carefully and check any key words you don't understand, it may help to build up your own glossary of terms.
- Seek help with any difficult ideas and key words in the article, particularly if they are barriers to your understanding of the article.
- Identify aspects of the A-level specification which are woven into the article and make sure you know the relevant A-level detail of material from the specification. Remember that you will need to draw on your own knowledge of the course as well as the content of the article to answer the questions. For example, see question (g) in the comprehension practice.
- Some background research into major items in the article may help your understanding.
- Your teacher does not know the questions and is therefore free to be a mentor in helping you to prepare for the exam.

This comprehension is based on the Unit 6 A2 Biology paper in June 2006. You will need to ask your teacher to find you a copy and read it carefully before you answer the questions.

Adapted from '*Life at the Extremes*' by Frances Ashcroft.

Published by Flamingo, 2001 ISBN 006551254.

Available from the Edexcel website www.edexcel.com.

Questions

The scientific article you have studied is adapted from a book called '*Life at the Extremes: the Science of Survival*' by Frances Ashcroft. Use the information from the article and your own knowledge to answer the following questions.

(a) Explain how humans are able to survive in dry air at temperatures above 100 °C.   (4)

(b) Inuits have evolved short stocky bodies. Suggest how this could have occurred. (2)

(c) Explain how sufferers of cystic fibrosis can be detected by a sweat test.   (2)

(d) Explain why marathon runners and cyclists are at a high risk of heatstroke at the end of a race.   (3)

(e) Explain how the study of pigs has led to a genetic test for malignant hyperthermia in humans.   (2)

(f) Suggest why taking aspirin may slow a person's recovery from an infection.   (3)

(g) Describe the role of calcium ions in muscle contraction.   (4)

(Total 20 marks)

Getting started with your investigation

You have to produce a report on an experimental investigation, which you have planned, carried out and interpreted individually. There are 45 marks available – 20% of the total A2 marks. The marks are awarded as follows: research & rationale (11 marks); planning (11 marks); observing & recording (8 marks); interpreting & evaluation (9 marks); communicating (6 marks).

Choosing your topic for investigation

ResultsPlus
Examiner tip

Choosing your own interesting question can often result in higher marks. However, discuss this with your teacher to make sure that it will be possible to collect sufficient data to meet all the criteria and that it can be carried out safely.

Do...	Don't...
investigate one variable only.	try to 'prove' anything or just try to demonstrate a well-documented 'fact'.
base your hypothesis on sound AS or A2 level biology and make it simple and clear.	repeat a core practical or a basic procedure from a book.
include a clear statistical statement.	do the same as your friends or classmates.
try to investigate an interesting question.	include lots of irrelevant material.
include clearly labelled subsections to match the criteria.	include illustrations that are not referred to or lots of graphs when one would do.
make sure you are really familiar with the marking criteria before you start.	forget to refer to the marking criteria when writing up your investigations.

ResultsPlus
Watch out!

Keep a careful check on your word count. It is very common for reports to be too long. This is often because research & rationale is not all relevant. Later sections tend to be rushed or too brief. As a rough guide use the mark allocations – so research & rationale would be about 700 words.

What is 'rationale'?

This is a reasoned explanation for your hypothesis. Research textbooks, magazines, scientific journals and the Internet but make sure that you only include material that is relevant to your hypothesis.

You must indicate clearly in your report exactly where you have used your researched information. (see Unit 3 Advice on referencing).

ResultsPlus
Build Better Answers

A good, clear hypothesis improves your chances of achieving a high mark in all the criteria.
Good hypotheses:
- 'There is a significant positive correlation between soil moisture content and the distribution of creeping buttercup (*Ranunculus repens*).'
- 'There is a significant difference between the rate of growth of pollen tubes from fresh pollen and pollen stored for 14 days.'
Poor hypotheses:
- 'Seaweeds on a rocky shore will show zonation.'
- 'Plax mouthwash will kill more bacteria than Listerine mouthwash.'

Planning

Your report must give clear evidence of how you have thought about and developed an effective and safe method to test your hypothesis. You need to use a simple trial experiment to do this.

Making sure you collect valid and reliable data

Always ensure you are really testing your chosen hypothesis.

- 'Make a clear plan of action including a trial experiment. This should explain what materials you intend to use and details of quantities and concentrations, etc. You also need to show exactly how you are going to use your trial to develop the final method.
- Decide how your data is to be analysed before data collection, not later. For example chose your statistical test and number of repeat measurements in advance.
- Include a separate section for discussion of possible variables and how you intend to control them or take them into account.

Considering variables

It is unlikely that you will be able to control every single variable but you must show that you have considered all the important factors that could affect your results and that your planned investigation will yield some scientifically meaningful data.

(a) In the laboratory

You should be able to control most variables here but even if apparatus is limited, use what is available, for example, you may not have a thermostatic water bath but you could use a beaker of water to hold temperature constant. Beware of making your own judgements of things like colour changes, try to devise a method that will help you to be consistent. Trial experiments are an ideal way to do this.

(b) In the field

Many variables are more difficult to control in the field but they still need careful consideration. Select your sampling sites carefully to make sure you do not introduce more variables. Measuring some abiotic factors can help to ensure they are not affecting the results.

Risk assessment

You must show that your investigation can be carried out safely by including a risk assessment. The diagram below shows one approach to this assessment.

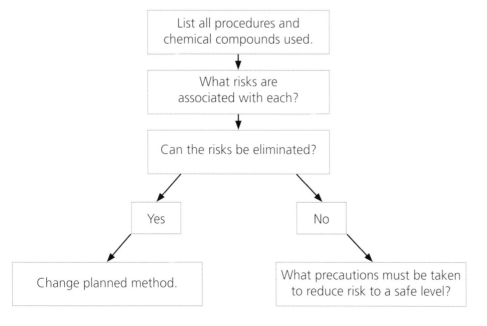

ResultsPlus
Watch out!

Poorly planned data collection can lead to the completion of many statistical tests or graphs which do not actually relate to the hypothesis you are testing. This can limit marks in planning and in interpreting & evaluation.

Planning – trial investigations

The following are good examples of how to use a trial experiment to change your method:

'The clear areas of agar were not circular so I decided to trace them on graph paper and measure the area by counting squares.'

'It is obvious from my table that there was only a very small difference in my results when I increased the concentration by 1% each time. I changed this to 5% each time so that the effect was much more obvious.'

'The stream I was using had lots of areas which were shaded by large trees and the depth of water was quite different in some parts. I searched further downstream until I found an unshaded section with an even depth but there was a pool with slow-flowing water and a narrower section with a faster moving stream.'

Don't write a detailed method twice. You can achieve high marks by writing this once then listing any amendments you make, with reasons, after your trial.

Make sure that your trial experiment is an important part of your planning, not just an exercise to justify what you have already decided. In order to gain high marks you will need to show that you have used the results of a trial to amend and develop your method. Simply following instructions or copying a common experiment from a book will only gain a few marks.

The trial does not have to be extensive but it should produce some evidence or simple data that you can use to explain any modifications you make.

If you are working in the laboratory you must try out your method and check to see if you can obtain the results you need. If you are working in the field you must visit the site and try out your sampling technique.

Some important questions for trial experiments

(a) In the laboratory

- Is your chosen range of values suitable?
- How accurately can you take measurements? Is there anything that can be done to make your measurements more accurate and reliable?
- Are there any variables that you have not taken into account?

(b) In the field

- Is the whole of the selected site suitable?
- Do you need to choose sampling sites to avoid introducing other variables?
- Can you identify the important species?
- Does your proposed sampling method allow you to collect sufficient data?

Observing and recording

Make sure you record your data in a clear table, using the correct SI units and at a level of accuracy that is sensible bearing in mind the methods you use.

Tabulation

- Use clear, accurate headings including units.
- Put units in the headings only, not with individual readings.
- Make sure there is a clear summary table showing the values used to draw your graph.

Significant figures

- Use a consistent number of significant figures for all data (including manipulated figures such as means).
- Remember to use zeros (2 is not the same as 2.00).
- Do not use numbers of significant figures that cannot be justified by your method. A simple way to apply this rule is not to use a greater number of significant figures than shown in your actual measurements.

Types of graph

There are a surprising number of basic errors in selecting and drawing graphs of a suitable type, so give this some careful thought, starting with basic principles:

- Normally there should be only one or two graphs.
- Your graph needs to be directly linked to your hypothesis and help you to make conclusions from your data.
- Axes should have clear, accurate labels with appropriate SI units where possible.
- Lines of 'best fit' are not a requirement at A2 level and should normally be avoided.
- Most simple line graphs are best presented by joining points accurately with a ruler.
- It is better to leave a scattergram without a line rather than draw one by guesswork.
- Avoid using 'sample number' as an axis. This is almost always meaningless when attempting to analyse trends and patterns in data and especially when random sampling has been used.

ResultsPlus
Watch out!

A very common mistake is to record time from a stopwatch as a decimal or as two units. Recording 2 min 50 s as 2.5 min instead of 2.83 min or 170 s is a very large error.

Presenting data to help analysis

If you are looking for significant differences in your investigation, you may have two sets of samples from which you would calculate two means. This might lead to a very simple two-column bar graph, which would give you only limited information on which to make detailed comments. However, selecting suitable size classes for each data set and then plotting a histogram for each on the same axes would allow you to comment on such things as the spread of data, any overlap and skewed distribution.

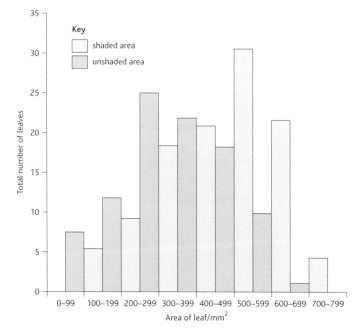

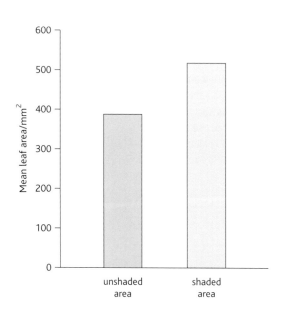

Dealing with data

Hypothesis testing using statistics

Imagine you are taking random samples from two areas and you wish to decide whether they are different. To make this decision scientifically you need to follow these rules:

1 Begin by assuming that the averages of the two samples are the same. This is a *null hypothesis*.

2 Take samples from each area.

3 It is very unlikely the averages will be identical so you can calculate the chance (probability) of obtaining results like yours, even if the two samples were not really different, using a statistical test.

4 If the chance (probability) calculated is higher than the *significance level* then you must accept that your first assumption is correct and there is no difference. If the chance of getting these results is lower than the *significance level* then you would reject the null hypothesis.

Examples of null hypotheses

- There is no significant difference in the numbers of mayfly nymphs found in slow-flowing streams and fast-flowing streams.
- There is no significant correlation between the abundance of creeping buttercup and soil moisture content.

Significance levels

In most investigations you should use a significance level of 5%. This means there are 5 chances in 100 that the results you obtain could occur even if there was no difference between the two sets of data. This can be written as a probability of p=0.05. To find this value for your data you normally have to calculate a test statistic and then look up the probability in a published table.

Statistical tests

You are not expected to know the formulae or fine details of each test. You should concentrate on selecting the correct type of test and demonstrating your understanding of how to interpret the results. The three types of test you are most likely to consider are:

- tests for significant difference e.g. t-test or Mann Whitney U test
- test for significant correlations e.g. Spearman's Rank Correlation test
- tests for significant association or 'goodness of fit' e.g. Chi Squared test

Other statistical terms

You are also expected to understand the following terms where applicable:

- *arithmetical mean (average)* – the sum of all the measurements divided by the number of measurements
- *median* – the middle value of your data where half the sample measurements are above this value and half are below
- *mode* – the measurement which occurs the greatest number of times in your sample.

ResultsPlus
Examiner tip

You cannot achieve more than 4/5 marks if you do not explain the results of your statistical test in your own words. You can use a computer programme to calculate the test statistic but not to explain its meaning.

Interpreting and evaluating

Conclusions

It is vital that you begin by describing accurately the trends and patterns shown by your data. At this stage do not be influenced by theoretical expectations. The next stage is to try to interpret your results using biological principles. This does not mean simply adding a lot of biological theory. You must use your biological knowledge and link it very clearly to your data.

ResultsPlus
Build Better Answers

Avoid terms such as 'proves that' or 'shows that'. Use more cautious terms such as 'supports the hypothesis that' or 'agrees with the suggestion that'.

Limitations

It is expected that you will take an objective, critical look at the method you have used and assess how it might affect the reliability of your conclusions. The important question to ask yourself is 'no matter how carefully I carry out this investigation what factors could still cause variations in my repeat readings?'

ResultsPlus
Build Better Answers

Poor limitations:
- admissions of practical incompetence when describing limitations
- suggestions of limitations that should have been eliminated by sensible planning or a trial experiment.

Good limitations:
- 'I set up a series of colour standards to make my judgement of the end-point as accurate as possible but this was still very subjective. It would be more accurate to filter my samples and use a colorimeter to give a precise measurement.'

Sensible modifications

This section should follow from your analysis of limitations. If you have identified factors which could cause variations then what modifications could be made to your method to minimise their effect?

Don't:
- simply suggest taking more samples – you should have thought about this in planning and just doing more of the same will not normally improve your method
- modify your investigation so that it begins to test a completely different hypothesis.

Communicating

High-scoring reports always make it clear where each of the criteria has been addressed by using sub-headings. This will prevent you omitting important details or repeating yourself. Use the advice given in other parts of this section of Unit 6 to tabulate your results accurately and choose the right form of graph.

Keep your sentences short but accurate and use a spelling check, especially for scientific terms. You will be given credit for selecting information from sources in 'research & rationale' but for higher marks you must use at least one professional journal and list all your sources in a bibliography, identifying where in your report each has been used.

For the highest marks you will need to evaluate your chosen sources, describing their credibility to scientists as a whole.

Answers to in-text questions

Unit 4: Topic 5

Photosynthesis 1 pp.8–9
QQ
1 reduced NADP; ATP
2 **a** gain of electrons; **b** loss of electrons

TT
1 The alga produces oxygen from the water it uses in photosynthesis, but only in the light. At all other times both types of oxygen, which are chemically indistinguishable, are being used in respiration therefore levels are falling due to this, both in the light and the dark. The fall in $^{16}O_2$ in the light is offset by its release from water.

Photosynthesis 2 pp.10–11
QQ
1 In recreating RuBP and in the formation of GALP. It is used in the 'first' step, carbon fixation.
2 Catalyses reaction of carbon dioxide with RuBP.

3

Structure	Function(s)	Features
thylakoid membrane	light-dependent reactions	contains chlorophyll, other accessory pigments, electron carriers
thylakoid space	photolysis of water	photolysis enzyme(s)
granum	provides a site for light-dependent reactions	large surface area
stroma	light-independent reactions	enzymes for all LIR stages including Rubisco
outer membrane	fully permeable	many open protein channels
inner membrane	permeable to many substances which need to enter or leave the chloroplast	gated and active transport channels, together with open channels

TT
1

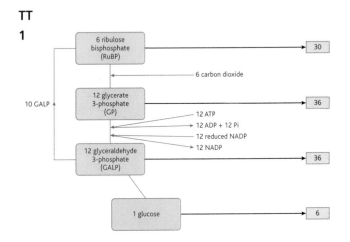

Energy transfer, abundance and distribution pp.12–13
QQ
1 R = GPP – NPP
2 The place where an organism lives and the role it plays (job it does) there.

TT
1 First need to work out the energy that is actually transferred to consumers. It is the total fixed minus respiration loss, i.e.
$1.9 \times 10^4 - 1.1 \times 10^4 = 0.8 \times 10^4$
so NPP $= 0.8 \times 10^4$
Production of primary consumers $= 0.1 \times 10^4$
so efficiency $= \dfrac{0.1 \times 10^4}{0.8 \times 10^4} \times 100 = 12.5\%$

Investigating numbers and distribution pp.14–15
QQ
1 a soil factor; any three of: pH, organic matter content, water content, texture, mineral content
2 Niche is a combination of what an organism does (e.g. its feeding type) and where it lives, i.e. where it does it – the 'address' and role of an organism is its niche.
Transect is a path along which occurrences of things are recorded (such as what plants grow there, what the pH is, etc.)
Quadrat is a fixed area sampling device. It is sometimes, but not always, gridded to make estimates, particularly of percentage cover, easier to work out.
3 e.g. Rocky shore: abiotic – salinity, temperature, light intensity/solar input; biotic – any three of: competitors, predators, parasites, herbivores. Sand dunes: abiotic – soil factors (as in Q1 – any three); biotic – any three of: competitors, predators, parasites, herbivores. Woodland: abiotic – soil factors (as in Q1 – any three); biotic – any three of: competitors, predators, parasites, herbivores.

TT

1

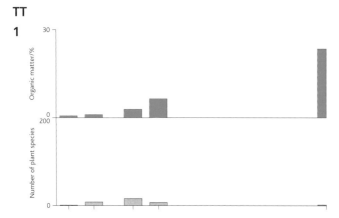

Species numbers reach a peak at 500 m from the reference point, whereas the organic matter content continues to rise throughout. There are few species in the first two quadrats because these are highly specialised pioneers in harsh conditions. Conditions improve with distance from the reference point (on the beach) as organic matter accumulates in the soil due to death of plants. Both mineral content and water-holding capacity also increase. This leads to an increase in species number, but this falls away again as a few species eventually dominate.

Speciation and evolution pp.16–17

QQ

1 reproductive isolation

2 Because the survivors must reproduce and pass on the favourable allele.

TT

1 Because they had only been separated for a relatively short time. (Or pupils may explain that evolution takes not just a few hundred years.) The chance mutations that might be beneficial have not happened. There is no significant (selection) pressure for change.

Greenhouse gases and the carbon cycle pp.18–19

QQ

1 E.g. tree ring data; ice core data; pollen analysis; temperature records.

2 Biofuels only release as much carbon as they have absorbed while they are growing, i.e. they are carbon neutral. The disadvantage is that they take up land which could be used for growing food, destroying habitats etc.

TT

1 Sketch of carbon cycle similar to the carbon cycle on page 26. It should show that human activity has increased carbon dioxide levels in the atmosphere by adding more carbon dioxide from burning fossil fuels and also decreasing the amount that plants and trees take in through deforestation.

Reforestation would increase CO_2 intake from the atmosphere by photosynthesis. Use of biofuels would reduce use of fossil fuels and are carbon neutral as the CO_2 released in combustion has only recently been fixed in photosynthesis.

Impacts of global warming pp.20–21

QQ

1 The temperature at which it catalyses its reactions most rapidly.

2 The study of seasonal events such as flowering times, hatching times, mating times, migration.

3 Although photosynthesis levels will increase it will be limited by other factors such as temperature or light levels. The increase in photosynthesis levels will not be sufficient to balance the increase in carbon dioxide levels.

TT

1 Must be handled properly (kept in water with at least 10 g of salt per litre and better to have 35 g salt /l); return to holding tank after inspection; use a soft plastic pipette to get them out of the water; make sure mouth of pipette wide enough; never observe for more than 5 minutes.

Unit 4: Topic 6

Decay and decomposition pp.26–27

QQ

1 Any three from: life cycle of an insect, state of decomposition, temperature of body, rigor mortis, succession of insects.

2 They break down biomass and release carbon dioxide into the atmosphere through respiration.

TT

1 In both there is a change to conditions that suit other species; in both there is a change in the species or distribution of species with time; in the sandy shore we see all stages at the same time; in the body we see only one stage.

DNA profiling pp.28–29

QQ

1 introns

2 Because the more STRs looked at, the more likely the profile is to be unique.

TT

1 DNA has fewer differences (in racehorses); DNA fingerprints are likely to be more similar; probability of DNA profile being unique much smaller; most horses closely related/humans are not

Answers to in-text questions

DNA and protein synthesis pp.30–31

QQ

1 intron
2 Each amino acid is coded for by more than one codon. Even if the last base of the codon is changed by mutation the correct amino acid will still be coded for.
3 molecule with anticodon at one end and amino acid at the other – tRNA
complementary copy of template strand – mRNA
DNA transcribed to mRNA – template strand

TT

1 DNA unwinds; mRNA strand assembled from coding strand of DNA; mRNA travels out of nucleus into cytoplasm; (post-transcriptional changes take place); mRNA associates with ribosome; tRNA molecules bring amino acids to ribosome; anticodon on tRNA associates with codon on mRNA and polypeptides bind together to make polypeptide chain; (polypeptide chain bound to other chains or other molecules to make protein).

Infectious diseases and the immune response pp.32–33

QQ

1 Similarities – any two from: both reproduce, both have protein, both nucleic acids, DNA and/or RNA, enzymes, organic molecules. Differences – any two from: bacteria have cell wall, capsule, flagellum, ribosome, pilus, mesosome, plasmid, cell membrane.
2 Because the high temperature might slow down bacterial reproduction as well as enhance the immune response and phagocytosis.

TT

B memory	B effector (plasma)	T helper	T killer	T memory
made from B cells after T helper activation	made from B cells after T helper activation	activated by antigen on phagocyte	destroy viruses or bacteria in infected body cells as well as the infected cell	help more rapid activation of immune system if same 'intruder' encountered in the future
help more rapid activation of immune system if same 'intruder' encountered in the future	produce antibodies	activate B cells		

Infection, prevention and control pp.34–35

QQ

1 Which antibiotic is most effective in controlling the population of a named bacterium?
2 Incubating below 30 °C discourages growth of human pathogens. Air is allowed into the Petri dishes to discourage the growth of anaerobes.

TT

1 Answers should include reference to four or more of the following points: rapid reproduction of the bacteria; there are many gene mutations or a relevant mutation is more likely in a given time; there is a wide variety of proteins; there is a great deal of variation in microbes; microbes are selected; there is selection for the resistant phenotype or gene; this does not take very long; but developing new drugs does take a long time.

Unit 5: Topic 7

Muscles and movement pp.44–45

QQ

1 Lactic acid/lactate builds up in fast-twitch fibres. Slow-twitch fibres have a good O_2 supply so they use aerobic respiration to produce ATP which does not produce lactate.
2 ATP is hydrolysed which causes the myosin head to change shape, ATP binding frees the myosin from the cross-bridge, ATP also used in active transport of calcium ions back in sarcoplasmic reticulum.
3 Muscles can only contract (shorten) so another muscle acting in the opposite direction is needed to extend the contracted muscle once it relaxes.

Energy and the role of ATP in respiration pp.46–47

QQ

1 Any four from: active transport, muscle contraction (sliding filament theory), glycolysis, Calvin cycle, protein synthesis, phosphorylation.
2 ATP is never stored, but can be rapidly produced or used in a cell; glycogen is a large, long-term store of chemical energy. ATP releases small quantities of useful energy when hydrolysed, glycogen contains a lot of stored energy that is not immediately available for the cell to use. ATP is soluble, glycogen is insoluble.
3 NAD is used to oxidise the triose phosphate (sugar) which produces ATP. Reduced NAD is used to reduce pyruvate into lactate so that glycolysis can continue.

TT

1 Similarities: phosphorylation involved, hydrogen carrier involved, ATP/glucose/triose phosphate/glycerate phosphate/eq involved; redox. Differences: NAD/NADP; ATP produced/needed; glucose used/produced; overall oxidation of glucose in respiration/reduction of CO_2 in photosynthesis; many more intermediates in Calvin cycle; cytoplasm/chloroplast.

Krebs cycle and the electron transport chain pp.48–49

QQ

1 carbon $\rightarrow CO_2$; hydrogen $\rightarrow H_2O$
2 Oxidative because the energy is transferred by redox reactions. Phosphorylation because ADP gains a phosphate group.
3 Without O_2 there will be nothing to accept the electrons (and hydrogen ions) at the end of the electron transport chain. The cell will run out of NAD and FAD so nothing can oxidise the compounds in the Krebs cycle (and link reaction).

TT

1 Glycolysis in cytoplasm, link reaction and Krebs cycle in matrix of mitochondria, oxidative phosphorylation/electron transport chain on cristae of mitochondria.

The heart, energy and exercise pp.50–51

QQ

1 $4.95\,dm^3$ per minute
2 Sketch should include repeated irregular QRS waves without many (or any visible) P and T waves.
3 Heart would continue to beat because it is myogenic, but would not change in response to changes in the body. However, it would still be able to respond to hormonal changes such as the release of adrenaline.

TT

1 Short-term effects: stroke volume and heart rate increase $\rightarrow$ cardiac output increases. Rate and depth of breathing (tidal volume) increases $\rightarrow$ ventilation rate increases. Long-term effects: stroke volume increases so resting heart rate decreases.

Homeostasis pp.52–53

QQ

1 A change in a factor brings about a response that counteracts the change so that the factor returns to a norm value.
2 Core body temperature will rise above 37°C; heat stroke; hypothalamus may become damaged; enzymes may be denatured; membrane proteins may be denatured; transport and respiration may be impaired; coma and/or death may result.
3 Heat gain centre of hypothalamus stimulates effectors; sweat production inhibited; reduced blood flow to the skin, shivering, metabolic rate (of liver) may

increase; hairs raised on skin; behavioural responses such as increased movement, put on extra clothes, etc.

TT

1 Some animals have lots of fast-twitch fibres for speed and power, e.g. predators to catch prey. Some animals may have lots of slow-twitch fibres, good breathing and circulatory systems (may be lightweight) for endurance, e.g. migratory birds who have to fly continuously for long distances.

Health, exercise and sport pp.54–55

QQ

1 Increased chances of obesity could lead to increased blood pressure $\rightarrow$ increased chance of damage to endothelium of arteries, higher concentration of low density lipoproteins $\rightarrow$ increased chance of plaque/atherosclerosis $\rightarrow$ coronary heart disease.
2 Less cytokines released $\rightarrow$ less activation of specific B and T killer cells $\rightarrow$ less chance of destroying pathogen before it can multiply enough to cause damage to tissues and create symptoms of the sore throat.
3 Transcription factors bind to the promoter site of the gene allowing the RNA polymerase to bind (transcription initiation complex) allowing transcription to take place so mRNA can be produced.

TT

1 Opinions will vary, but your answer may consider other issues such as sports equipment, funding and time for training, coaching, altitude training, and diet. You should give your opinion and justify your answer.

Unit 5: Topic 8

Responding to the environment pp.60–61

QQ

1 Photoreceptors detect light.
2 Positive phototropism in shoots enables them to grow towards the light. Roots are negatively phototrophic to help them grow down into the soil away from the light.
3 cell elongation

TT

1 The two systems working together provide greater coordination and control between short-term and long-term responses and changes. Nervous system is fast for immediate response, endocrine system includes control of growth and development.

Answers to in-text questions

Neurones and nerve impulses pp.62–64

QQ

1 Depolarisation makes the p.d. across the membrane less negative (because positive sodium ions move into the cell), whereas with hyperpolarisation the p.d. becomes more negative (e.g. if negative chloride ions move into a cell at an inhibitory synapse).
2 The refractory period means that the sodium ion channels can not reopen which prevents another action potential being triggered. Synapses only have receptors on the postsynaptic membrane.
3 Voltage-gated sodium ion channels open and sodium ions diffuse into the cell.

TT

1 Sodium ions move in during action potential, potassium ions move out. During recovery, sodium ions pumped out and potassium ions in. Potassium ions diffuse out during resting potential/state. Other labels could include passive diffusion, facilitated diffusion, active transport, osmosis, endocytosis, exocytosis, etc. with suitable examples.

Vision p.65

QQ

1 In the dark rod, cells remain slightly depolarised because of the open sodium ion channels and continually release the inhibitory neurotransmitter.
2 breaks down into opsin and retinal.
3 Photoreceptors in plants are chemicals like phytochrome, in mammals they are specialised cells like rods.

TT

1 Rods are more sensitive than cones but do not detect differences in colour. Spatial summation at the bipolar cell – three rod cells converging on a single bipolar cell have a greater chance of affecting the biopolar cell in low light levels than a single cone cell.

The structure of the human brain p.66

QQ

1 The cerebellum is concerned with the control of balance and movement, the cerebrum is involved in the ability to think, see, learn and feel emotions, etc.
2 the frontal lobes of the cerebrum (cerebral hemispheres)

Brain development pp.67–68

QQ

1 The child may become blind because the eye was deprived of light during the critical window.
2 Kittens and monkeys have a similar visual system and brain to humans so they are good models for looking at human brain development. You would not be allowed to do the same experiments on human children.
3 Schizophrenia is mainly determined by genes, but there is probably a small environmental influence.

TT

1 There is no correct answer, but your opinion should be justified with reference to potential benefits of developing the new drugs, requirement for animal tests before human trials, animal welfare/rights issues.

Learning and habituation p.69

QQ

1 stimulus – jet of water → pressure receptor on siphon → sensory neurone → motor neurone → gill muscle → response – gill withdraws
2 Sea slugs raised in the sea may already be habituated due to water currents and tidal changes in the sea.
3 Genetic programming (nature) is likely to be responsible for innate reflexes, because they are not influenced by the environment (nurture).

Effects of imbalances in brain chemicals pp.70–71

QQ

1 Dopamine is active in the part of the brain that deals with emotions. The other symptoms of Parkinson's and the knowledge that there is no cure would also leave many people feeling depressed.
2 L-dopa can pass into the brain and be made into dopamine.
3 The new drug may bind to serotonin receptors and mimic the effect of serotonin causing action potentials to form.

TT

1 Delivery of drugs to the specific area of the brain in the correct concentration and released at the precise time needed difficult/impossible.

Uses of genetic modification pp.72–73

QQ

1 Genes may be transferred into wild species.
2 Can be produced cheaply, in bulk and easily purified. The human protein is produced rather than using an animal protein so there should be fewer side effects.

TT

1 Benefits: early treatment/preventative care; carriers can choose to avoid having children/use embryo/fetal screening.
Disadvantages: stress for those with gene; cost; some couples may choose not to have children; insurance issues; eugenics issues.

Answers to practice questions

Unit 4: Topic 5

1 (a) 1. thylakoid/granum;
   2. membrane;   2
(b) A ATP;
   B reduced NADP/eq;   2
(c) photolysis;   1
(d) 1. less carbohydrate production;
   2. less reduced NADP;
   3. less reduction of carbon dioxide;
   4. less ATP (to supply energy);
   5. less conversion of GP to GALP;   4
   Total 9 marks

2 (a) Measure {growth/height/number of leaves/mass/
   dry mass};
   Growth with copper and {without copper/control/
   range of copper concentrations};
   Reference to controlling variables;
   Reference to {repeats/means/calculation of
   percentage growth};   2
(b) Plants compete/eq;
   For {ions/water/nutrients/nitrates/tight/space/eq}
   Tolerant plants less well adapted/converse/eq
   So tolerant plants {smaller/less dry mass}/
   converse/eq
   Idea: Results are due to competition only because
   trays 1 and 3 growing the same;
   Manipulated figures;   4
(c) Decreases/more non-tolerant;
   No benefit;
   Competes less well;   2
   Total 8 marks

3 (a) (i) GPP and NPP similar to start with;
   both increase;
   (after 2 days) GPP and NPP diverge/eq;
   figures in support;   2
   (ii) more energy is used in metabolism/eq in
   {older/bigger} plants;
   figures in support;
   suitable explanation, e.g. protein synthesis/
   flower initiation/differentiation/ref. to
   herbivores;
   more photosynthesis tissue;
   (as grows)/eq;   2
(b) GPP–NPP=R/eq;
   biomass production reduced by
   respiration/eq;   2
   Total 6 marks

4 (a) proportion of total alleles;
   for one gene (in a population)/eq;   2
(b) different alleles exist/ref. mutation;
   advantage in specific environment;
   ref. selection pressure;
   more likely to reproduce;
   allele passed to offspring more often;

ref. at disadvantage in other environment;
(Allow converse argument)   4
(c) faster life cycle of bacteria/converse/eq;
   greater selection pressures on bacteria (e.g.
   antibiotic use);
   ref. plasmid transfer in bacteria/eq;
   larger numbers of bacteria hence larger gene
   pool/eq;
   ref. mutation;   2
   Total 8 marks

Unit 4: Topic 6

1 (a) 1. C is bactericidal;
   2. bactericidal kills bacteria;
   3. B is bacteriostatic;
   4. bacteriostatic prevents reproduction/growth; 3
(b) 1. bacterium is no longer affected by antibiotic A;
   2. reference to mutation/changed {gene /DNA};
   3. reference to resistance;
   4. reference to selection/eq;
   5. reference to plasmid transmission/horizontal
   inheritance;   4
(c) 1. lawn bacteria/eq;
   2. reference agar plate/eq;
   3. antibiotic in well/multidisc/eq;
   4. incubation qualified;
   5. measurement of clear area/eq;
   6. bigger area implies more effective;
   7. reference to safety/aseptic technique/eq;   4
   Total 11 marks

2 (a) 1. similar route for infection;
   2/3.examples of means of transmission;
   4. immunosuppression in HIV/eq;
   5. reference to opportunistic infection/eq;   2
(b) T(-cell)/T-lymphocyte/T-killer;   1
(c) 1. signal from surface protein;
   2. activation of PKR;
   3. ref. to protein synthesis/translation/eq;
   4. no production of virus;
   5. cell death/cell function disrupted: 3
(d) 1. rapid reproduction.
   2. many (gene) mutations/relevant mutation
   more likely in given time;
   3. ref. to variety of proteins;
   4. large variation;
   5. selection;
   6. of resistant phenotype/gene;
   7. short time;
   8. long drug development time;   2
   Total 8 marks

3 (a) 1. T helper cells {destroyed/damaged/reduced in
   number/cell lysis/eq};
   2. no T killer cell {production/activation}/eq;
   3. B cells activation/plasma cells production/eq;
   4. (less/no) antibody production/eq;
   5. phagocytosis/phagocytes;   4

(b) 1. (inflammation) – preventing infection at site of tissue damage/detail of response e.g. macrophages attracted/oedema/increased blood flow;
2. phagocytosis;
3. (lysozyme action – enzyme to) destroy bacteria/cell lysis/breakdown of cell walls;
4. interferon; 4

Total 8 marks

4 (a) 1. insects development rate/life cycle/aspect of development/rate of growth/eq;
2. temperature dependent;
3. rate fixed at constant temperature;
4. ref. to weather conditions affecting temperature of the room; 2

(b) several values for development rate/eq; 1

(c) 1. Do not know temperature throughout/ temperature not constant;
2. so rate might vary;
3. Might have died at night;
4. so infestation next morning/do not know time of infestation/entry to room difficult;
5. Too late for other evidence;
6. Other factor involved; 2

Total 5 marks

Unit 5: Topic 7

1 (a) medulla (oblongata); 1

(b) 1. ref. to increase in rate of (anaerobic/aerobic) respiration;
2. {increase in carbon dioxide levels/increase in {lactic acid/lactate} levels/decrease of pH/ increase in {hydrogen ions/H$^+$}/increase in carbonic acid} in the **blood**;
3. ref. to chemoreceptors;
4. {aortic/carotid} bodies or ref. to carotid {artery/ aorta};
5. {cardiac/cardiovascular} centre (in medulla);
6. **{more/increase in frequency}** of impulses along sympathetic nerve;
7. ref. to SAN/sinoatrial node; 4

Total 5 marks

2 *Immune suppression:*
1. fewer infections/immune system most effective with moderate exercise
or {too little/too much} exercise {suppresses immune system/more infections};
2. fewer natural killer cells;
3. details of changes (to cells) of the immune system after vigorous exercise;
4. inflammatory response in muscles reduces (non-specific) immune response elsewhere;
Joint damage:
5. (articular) cartilage/patella/bursae/ligaments/ muscle (fibre);
6. further details of damage;
7. detail of how damage occurs;

Benefits to cardiovascular system:
(max. 4 marks from this section)
8. moderate exercise lowers (resting) blood pressure;
9. due to increased arterial vasodilation;
10. improves ratio of HDL to LDL/increases levels of HDL/reduces levels of LDL;
11. beneficial effects of exercise on obesity/BMI;
12. ref. to prevention (type II) diabetes;
(max. 6 marks) 6

Total 6 marks

3

Stage	Part of cell in which it occurs	Two products
	cytoplasm/cytosol	pyruvate/pyruvic acid, NADH and ATP
		ATP, CO_2, NADH and FADH$_2$ /H$^+$
	cristae/inner membrane of mitochondrion	

4

Total 4 marks

4 (a) 1. reading 22 and 15;
2. correct numerator :{7/22-15) x 100%;
3. correct denominator: 22 = correct answer; [correct answer 28.57: - 36.36%] 3

(b) 1. ref to Ca^{2+} binding to muscle protein (troponin);
2. (this) results in {change in shape of muscle protein (troponin)/movement of muscle protein (tropomyosin)/exposure of (myosin) binding site on actin};
3. ATP binds to myosin (head);
4. myosin attaches to actin (binding site);
5. ATP broken down to ADP and Pi/energy released from ATP/release of ADP and Pi;
6. head swivels/changes angle/idea of actin moving (over myosin); 4

Total 7 marks

5 (a) medulla/medulla oblongata; 1

(b) before diving/singing/sniffing/playing a musical instrument/peak flow measurement/ breathalyser/eq; 1

(c) 1. ref to baroreceptors/stretch receptors;
2. detect degree of stretch in diaphragm and/or intercostals muscles;
3. (and) feedback this to the {medulla / respiratory control centres} along (sensory) neurones;
4. idea of nervous link from {medulla/respiratory control centres}to muscles to {alter/change} {ventilation (rate)/breathing rate/(depth of) breathing}; 3

(d) 1. idea that exercise {increases the {CO_2/H$^+$} concentration/{decreases the pH/O_2 concentration};
2. (this increase is) detected by chemoreceptors;
3. (chemoreceptors) send impulse to medulla;

4. (resulting in) increased {rate of ventilation/breathing rate/depth of breathing/contraction of {respiratory/intercostals/diaphragm muscles}}; 2
Total 7 marks

6 (a) (i) B
    (ii) A (muscle opposite); 1
(b) (Only a small cut) because damage is less/less bleeding/less pain;
Recovery is rapid/shorter stay in hospital/eq;
Less risk of infection/inflammation; 2
(c) ref. to **pathogens**/disease causing organism;
through {travel/team sports/idea of runners meeting from other areas}/eq; weakened/suppressed immunity (with hard exercise);
through fall in natural killer cells/phagocytes/lymphocytes/T-helper cells/B and T cells;
Reference to airborne/droplet infection; 3
Total 6 marks

7 (a) sinoatrial/SA node /SAN /pacemaker; 1
(b) Any four from:
    1. wave of electrical impulses/depolarisation from SA node;
    2. passes over both atria;
    3. resulting in atrial systole;
    4. slight delay at AV node;
    5. Impulses pass along bundle of His;
    6. along Purkyne fibres;
    7. correct direction of impulse described/ventricles contract from the base up;
    8. resulting in ventricular systole; 4
Total 5 marks

Unit 5: Topic 8

1 (a) (i) 1. (light hits) photoreceptors (on the retina);
    2. impulses pass to the brain;
    3. ref. to sensory neurone;
    4. ref. to innate/inborn/autonomic response;
    5. impulses along parasympathetic nerve;
    6. ref to motor neurone;
    7. circular muscles contract/radial muscles relax;
    8. pupil {contracts /constricts/becomes smaller}; 4
    (ii) 1. {faster/eq} impulses due to;
    2. myelin acting as an {electrical/ eq} insulator;
    3. ref to Schwann cells producing myelin;
    4. depolarisation only occurs at the nodes of Ranvier;
    5. ref. to saltatory conduction;
    6. need rapid response to protect retina; 3
(b) 1. visual stimulation is essential for visual development;
    2. ref. to critical window/critical period/sensitive period;
    3. ref. to visual cortex;
    4. growth of axons/formation of synapses/inactive synapses eliminated;

---

    5. kittens less than 4 weeks old have not developed (visual cortex) {connections synapses} *or* kittens over 5 weeks old have already developed (visual cortex) {connections/synapses} 3
(c) 1. ref. to animal experiments helping to test {medicines/treatments}/give greater understanding of the {human/animal} body;
    2. ref. to utilitarian philosophy;
    3. expected benefits greater than expected harms/eq;
    4. reduces chances of harm when testing on people; 2
Total 12 marks

2 (a) 1. (rods contain) rhodopsin;
    2. ref. to convergence/summation/eq;
    3. therefore the dog will have better {vision in dim light/night vision}/eq;
    4. idea that dog can look directly at object (in dark)/eq;
    5. dogs are {more active at night/nocturnal}/eq; 3
(b) 1. ref. to phytochromes;
    2. name two forms {PFR and PR/P730 and P660};
    3. ref. to absorption of light (by phytochromes);
    4. conversion of PR to PFR **AND** reference to red light;
    5. conversion of PFR to PR **AND** reference to far red light; 3
Total 6 marks

3 (a) B (cerebellum); 1
(b) C (medulla); 1
(c) A (cerebrum); 1
Total 3 marks

4 (a) (characteristics) controlled by more than one gene/many genes; 1
(b) 100%/above 95% suitably qualified reference to mutations; 1
(c) An explanation to include two from:
    1. multifactorial;
    2. environment has an effect/genes and environment (interact);
    3. correct use of figures; 2
Total 4 marks

5 (a) the total of all the {genes/genetic material/DNA/alleles} in {humans/an organism 1
(b) (i) 1. ref. to example, e.g. avoid smoking/eat a special diet/avoid fatty foods/ more exercise;
    2. the need to be particularly careful when one knows one is particularly at risk;
    3. could have treatment in advance of onset of condition/ could make preparation for coping with problem/more check ups/closer monitoring; 2
    (ii) 1. to inform health service {planning/budget/priorities}/identify people at risk;
    2. early {treatment/diagnosis} {may reduce problems later/may help determine the appropriate dose of medication};

3. advising people with defective genes about having children/deciding whether to abort affected fetuses;
4. thus reducing cost /burden to society;
5. reduces insurance premiums for people without genetic defects;
6. to determine medical research priorities;   2

(iii)   Any two from:
1. undue intrusion into people's lives by government/infringement of {civil liberty/ human rights};
2. easier to cope if you don't know in advance/prefer not to know/creates needless stress;
3. implication for insurance premiums;
4. risk of discrimination;
5. pressure to have abortions/to avoid having children;
6. cost too much/too much taxation;

Any one development mark in the context of one of the points above from:
7. the benefits are not worth the risks /costs;
8. {health/life} insurance too expensive for those at risk/people might have to declare results of screening to get insurance;
9. lack of confidence in {government/ administrators} to keep data confidential/ data protection issues;   3

Total 8 marks

Answer to specimen paper questions

Unit 4

1 (a) X – chlorophyll/eq;
Y – NADP/ NADP+/eq;   2

(b) 1. provides {H/electrons/eq};
2. reference to reduction;
(Accept reducing power, reducing agent)
3. GP is (reduced)/ eq;
4. to produce GALP/eq;
(Allow mpts as they can be applied to their incorrect reduced Y)
5. during light-independent {stage/eq}/Calvin cycle/eq;   3

(c) (i) 1. electrons return to chlorophyll (so there is fluorescence);
2. carriers can not {accept/take up/eq} electrons;
3. because electron carriers are reduced;   2

(ii) carriers available again/electrons passed to NADP/eq;
(Accept carriers have passed as electrons idea)   1

(d) 1. NADP stays reduced /NADPH not used/eq;
2. because no GP {formed/available}/eq;
3. because RuBP cannot take up CO_2/eq;
4. because rubisco inactivated/eq;
5. (because NADPH not used) all electron carriers reduced/path B blocked/eq;
6. (therefore more) electrons {return to chlorophyll/follow path A/eq};   4
Total 12 marks

2 (a) 1. tree types can be identified from their pollen;
2. pollen only produced by {fully-grown/mature/eq} trees;
3. trees need to {grow/eq} for a long time before maturity/eq;   2

(b) (i) D   1
(ii) 1. reference to distributed across several climatic zones;
2. reference to little fluctuation in the pollen data from different ages;   2

(c) 1. climate has become warmer/eq;
2. reference to a change between 8700 and 6390 years ago;
3. {larch/spruce} were growing but died out/eq;
4. (larch/spruce) are only found in boreal and northern temperate regions (in present day);
5. (boreal and temperate regions) are cold climates;
6. pine was not growing but has become established more recently;
7. pine is only found in southern boreal and temperate regions (in present day);
8. (southern boreal and temperate regions) are warmer climates;   5

(d) Award one mark for each of the following points.
1. idea that dendrochronology uses evidence from {tree/annual} rings;

2. {density/thickness/eq} of rings changes with climatic conditions/thicker ring indicates warmer year;   2
Total 12 marks

3 (a)

Structural feature	Bacteria	Viruses
mesosomes	✗	
capsid		✗
nucleic acid	✗	✗
cytoplasm	✗	
ribosomes	✗	

   5

(b) (i) 1. Increase in number of new cases in Africa and Europe;
2. Decrease in number of new cases in Asia and South America;
3. Any relevant manipulation of data;   2

(ii) More incidence of TB in the population/eq;
(Award one mark for each of the following points in context to a maximum of *two* marks.)
1. Ref. to opportunistic infection;
2. HIV positive people have weakened immune system;
3. A higher proportion of HIV positive people are infected by TB;   3

(c) Two from:
1. TB bacteria {mutate/become resistant to antibiotics};
2. immigration from countries with high incidence of TB;
3. increased travel;
4. increase in HIV infection;
5. lower rates of immunisation against TB;   2
Total 12 marks

4 1. {nutrient/eq} agar plate;
(ignore Petri dish)
2. {lawned/inoculated/spread/eq} with bacterium/eq;
3. application of sample of antibiotic described
4. incubation described;
5. ref clear/inhibition zone;
(Max. 4 if no safe working)
Safe working –
6. ref aseptic technique/aspect of;
7. Petri dish not completely sealed;
8. low temperature of incubation below 30 °C;   5
Total 5 marks

5 (a) 1. fixed/constant area;
2. reference to sampling;
3. valid comparisons possible;
4. easy so can be repeated;   2

(b) 1. sampling along changing conditions/ environmental gradient;
2. systematic sampling /random sampling does not show distribution/eq;   2

(c) 1. more coverage by plants in 5/converse;
2. more organic matter in 5/converse;
3. more species in 5/converse;
4. different species present;
5. credit figures (e.g. 3 times more plants, 2.4 g more matter, 17 more species); 3

(d) 1. different communities at different distances;
2. few species near beach;
3. reference to pioneer species;
4. organic matter (increase with distance from beach);
5. consequence of increased organic matter (e.g. increased water holding, mineral content);
6. suited to more species further from beach;
7. reference to competition;
8. few dominant species;
9. (might be) climax community/mature community; 5

Total 12 marks

Unit 5

1 (a) 1. {calcium ions/Ca^{2+}} released from sarcoplasmic reticulum;
2. calcium (ions) binds to troponin;
3. (troponin) causes tropomyosin to move;
4. exposing (myosin) binding sites (on actin);
5. myosin head attaches to binding site/cross bridge formation;
6. myosin head {moves/nods forward/eq};
7. release of ADP and inorganic phosphate;
8. actin slides over the myosin;
9. (ATP causes) myosin head to detach;
10. {ATP hydrolysis/ATPase}; 5

(b) 1. ref. to prevention of release of neurotransmitter from presynaptic membrane;
2. similar shape (to neurotransmitter);
3. {binds/blocks/fits into} receptor on postsynaptic membrane;
4. ref. to {sodium ion/Na^+/cation} channels/hyperpolarisation/permanent depolarisation} of postsynaptic membrane;
5. no nerve impulses/action potentials/continuous action potential/eq;
6. inhibits acetylcholinesterase/breakdown enzyme/(bungarotoxin) not affected by breakdown enzyme; 3

Total 8 marks

2 (a) (i) 1. reference to oxygen (concentration) decreasing/eq;
2. greater (decrease) when ADP is added;
3. (oxygen used to) convert ADP to ATP (in respiration);
4. oxygen is needed for respiration/eq;
5. correct reference to oxidative phosphorylation;
6. reference to {ADP concentration/eq} is limiting; 3
(ii) reduced NAD/NADH/NADH2; 1
(iii) cristae/inner membrane/stalked particle; 1
(iv) 1. hydrogen atoms split into protons and electrons/eq;

2. electrons transferred along electron carriers/a series of redox reactions/eq;
3. oxygen is the terminal electron acceptor/water is formed;
4. {protons/eq} moved into intermembrane space/eq;
5. {protons/eq} move (into matrix) down a {concentration/electrochemical} gradient;
6. through stalked particles/ATP synthetase/eq;
7. correct ref. to chemiosmotic theory;
8. (ATP)formed by {phosphorylation of ADP/oxidative phosphorylation }/eq; 4

(b) (i) 1. correct reference to ATP (supplies energy) for active transport/reference to sodium–potassium pump/eq;
2. sodium ions pumped out (of the axon)/restores (membrane to) resting potential; 2
(ii) 1. correct reference to ATP (supplies energy) for active transport/reference to sodium–potassium pump/eq;
2. (pumps sodium ions out) of inner segment/maintains (more) negative charge inside the membrane/eq; 2

Total 13 marks

3 (a) 1. to map (human) chromosomes/to find where each gene is located on (human) chromosomes;
2. to determine base sequence;
3. international project (about human genes); 2
(b) (i) 1. to warn people at risk to take precautions/make lifestyle changes;
2. to plan medical provision (for the individual);
3. to determine NHS priorities/eq;
4. to warn people when there is a risk that if they have children they may have genetic disorders;
5. will make it easier to develop {ways of treating genetic deficiencies/gene therapy}; 2
(ii) 1. may mean people put under (undue) pressure to {have abortions/not have children};
2. may lead to discrimination over {jobs/insurance premiums};
3. may lead to (other ethically questionable) developments such as 'designer babies'/eugenics/immigration;
4. {stress/anxiety} – knowing something might happen may cause psychological stress even if it never happens in your lifetime/people may not believe test is reliable/people may not want to know;
5. {civil rights/personal freedom} – who should decide who should have genetic tests?/who decides who deserves very expensive treatment on the NHS?;
6. data protection issues – who will have access to genetic information about individuals; [this marking point could be a development of marking point 2] 3

Total 7 marks

4 (a) 1. evaporation of water (in sweat);
2. (evaporation) has a cooling effect / eq;
3. appropriate {reference to / description of} latent heat; (2)

(b) (i) 1. temperature dropped {from 0 to 15 minutes / when in the bath};
2. increased {from 15 to 25 minutes / when sitting on the chair};
3. lowest {at 15 minutes / when 'he got out of bath'};
4. credit a manipulated change in temperature; (3)

(b) (ii) 5 to 10 minutes:
1. temperature of water lower than body temperature / eq;
2. heat lost by conduction (to water); 15 to 25 minutes:
3. increased metabolism / shivering / eq;
4. generates heat / eq; (3)

Total 8 marks

5 (a) 1. calculation; 2. answer (= 92.9%); (2)
(b) 1. (as cycling speed increases) more carbon dioxide produced;
2. {carbon dioxide / low pH} stimulates breathing / eq;
3. increased need for oxygen / eq; (2)

Total 4 marks

6 (a) (i) pyruvate/pyruvic acid;   1
(ii) 1. (stage) 1;
2. (stage) 3;   2
(b) (i) a {series/sequence/eq} of (chemical) reactions/ each step is controlled by an enzyme/product of one reaction is the substrate for the next/eq;   1
(ii) matrix of a mitochondrion;   1
(c) (stages) B, C, D (and) F;   1

Total 6 marks

7 (a) (i) values between 0.4 to 0.55 × 12; = values between 4.8 and 6.6 dm^3 min^{-1};   2
(ii) values between 1.1 and 1.3 × 36 = values between 39.6 and 46.8 dm^3 min^{-1}; increased by about 6 times/increase of between 33.0 and 42.0;   2
(b) (i) heart rate × stroke volume or volume of blood pumped out of the heart in 1 minute. 1
(ii) As the minute volume increases the tidal volume (volume of oxygen breathed in) increases; increased diffusion of oxygen into blood (or muscle); increase in cardiac output increases volume of oxygenated blood reaching muscles;   2

Total 7 marks

Unit 5: Comprehension practice model answers

a Temperature receptors in the skin and hypothalamus detect the rise in temperature (1) and cause an increase in the volume of sweat produced (1). The sweat evaporates from the surface of the skin taking heat energy away from the body (1). This cooling can continue as long as the person is able to replace the water and salt lost due to the increased sweating (1).

(Max. 4 marks)

b Inuit with short stocky bodies are at a selective advantage (1) because they have a lower surface area to volume ratio (1) and will therefore lose less heat to their surroundings.

(Max. 2 marks)

c Sufferers of cystic fibrosis have a CFTR protein channel does not work (1). As a result less water moves from cells into sweat glands (1) so the sweat ends up with a higher concentration of salt than normal that can be detected in the sweat test (1).

(Max. 2 marks)

d Marathon runners will generate a lot of heat during the race because of the high rate of respiration (1). However, at the end of the race there will be less air flow over the body (1) so less sweat may evaporate (1). This may cause the core temperature to rise (1) resulting in a heat stroke.

(Max. 3 marks)

e Porcine stress syndrome was noticeably similar to malignant hyperthermia (1). They were therefore able to identify the human gene for malignant hyperthermia through comparison to the identified gene for porcine stress syndrome (2).

(Max. 2 marks)

f Aspirin blocks the synthesis of prostaglandins (1) and therefore reduces fever in the body (1). This could be a problem because a rise in body temperature may help to kill bacteria (1) and increase the activity of macrophages (1) in the non-specific immune system, reducing the effectiveness of the body's response to the infection.

(Max. 3 marks)

g Ca^{2+} ions are released into the sarcoplasm (1) from the sarcoplamic reticulum (1) in response to a nerve impulse arriving at the neuromuscular junction (1). Ca^{2+} ions attach to troponin (1) causing tropomyosin to move, exposing myosin binding sites on the actin filaments (1). This allows myosin to join to actin (1) starting the contraction of the muscle.

(Max. 4 marks.)

Index